MONOGRAPHS ON
STATISTICS AND APPLIED PROBABILITY

General Editors

D.R. Cox, D.V. Hinkley, N. Reid, D.B. Rubin and B.W. Silverman

(Full details concerning this series are available from the Publishers.)

Predictive Inference: An Introduction

SEYMOUR GEISSER
School of Statistics, University of Minnesota

CRC Press
Taylor & Francis Group
Boca Raton London New York

CRC Press is an imprint of the
Taylor & Francis Group, an **informa** business
A CHAPMAN & HALL BOOK

First published 1993 by Chapman & Hall

Published 2019 by CRC Press
Taylor & Francis Group
6000 Broken Sound Parkway NW, Suite 300
Boca Raton, FL 33487-2742

© 1993 by Taylor & Francis Group, LLC
CRC Press is an imprint of Taylor & Francis Group, an Informa business

First issued in paperback 2019

No claim to original U.S. Government works

ISBN 13: 978-0-367-44991-9 (pbk)
ISBN 13: 978-0-412-03471-8 (hbk)

**Visit the Taylor & Francis Web site at
http://www.taylorandfrancis.com**

**and the CRC Press Web site at
http://www.crcpress.com**

Library of Congress Cataloging in Publication Data

Geisser, Seymour.
   Predictive inference: an introduction/by Seymour Geisser.
       p.   cm.—(Monographs on statistics and applied probability: 55)
   Includes bibliographical references and index.
   ISBN 0-412-03471-9
   1. Prediction theory.   2. Analysis of variance.   I. Title
II. Series.
QA279.2.G45   1993   92-39408
519.2′87—dc20   CIP

Dedicated to my parents

Leon Geisser 1902–1977
Rose Geisser 1902

Contents

Preface

Prediction was the earliest and most prevalent form of statistical inference. This emphasis changed during the beginning of this century when the mathematical foundations of modern statistics emerged. Important issues such as sampling from well-defined statistical models and clarification between statistics and parameters began to dominate the attention of statisticians. This resulted in a major shift of emphasis to parametric estimation and testing.

The purpose of this book is to attempt to correct this emphasis. The principal intent is to revive the primary purpose of statistical endeavor, namely inferring about realizable values not observed based on values that were observed.

The book is addressed mainly to statisticians and students in statistics who have had at least a year of theoretical statistics that required knowledge of linear or matrix algebra and advanced calculus. At least a passing familiarity with the Bayesian approach is also helpful. Research workers in substantive areas of science or technology who rely on statistical methods and are cognizant of statistical theory should also be attracted to statistical methods that stress observables rather than parameters.

Both traditional frequentist and other nontraditional predictive approaches are surveyed in a single chapter. All subsequent chapters are devoted mainly to Bayesian predictive methods or modifications thereof. They are often introduced with a brief discussion of the purpose for which data are to be used. A theoretical apparatus is then delineated. This is exemplified using a variety of common statistical paradigms. Some of these paradigms are illustrated numerically to give the reader a firmer grasp of their execution.

Most readers will be familiar with the more customary concerns of statistical inference such as comparing either various treatments on a population or responses from different populations and more generally regression analysis. These are dealt with here in a predictive mode. Once the ideas are presented, the reader will think of other applications of predictive inference within this standard genre. More emphasis is placed on topics not usually associated with an introduction, such as model selection, discordancy, perturbation analysis, classification, regulation, screening, and interim analysis.

There are many more areas for the utilization of predictive methods than are presented in this monograph. These will occur to the reader once the ideas and direction presented are internalized.

Some familiarity with multivariate normal analysis and linear models is necessary for reading Chapter 9 and Example 10.8 of Chapter 10. Other than a modest acquaintance with the above cited topics, the mathematical and statistical background, previously indicated, should be sufficient.

I am indebted to my graduate students Li-Shya Chen Lee and George D. Papandonatos for reading and correcting errors in the manuscript and particularly with computing assistance. I would also like to acknowledge the Lady Davis Trust that helped make it possible for me to spend several months on the project at the Hebrew University of Jerusalem, and to the National Institutes of Health for research support.

CHAPTER 1

Introduction

Currently most statistical analyses generally involve inferences or decisions about parameters or indexes of statistical distributions. It is the view of this author that analyzers of data would better serve their clients if inferences and decisions were couched in an observabilistic or predictivistic framework. The focus on parametric statistical analysis arose from the measurement error model. Here some real physical entity was to be measured but the measuring instrument was subject to error. Hence the observed value was modeled as

$$X = \theta + e$$

where θ was the true value and e represented a normally distributed error of measurement. This measurement error convention, often assumed to be justifiable from central limit theorem considerations and/or experience, was seized on and indiscriminately adopted for situations where its application was dubious at best and erroneous at worst. Constituting the bulk of practice, applications of this sort regularly occur in technology but are much more frequent in the softer sciences. The variation here is rarely of the measurement error variety. As a true physical description the statistical model used is often inappropriate if we stress hypothesis testing and estimation of the "true" entities, the parameters. If these and other such models are considered in their proper context, then they are potentially very useful, i.e., their appropriate use is as models that can yield adequate approximations for the prediction of further observables presumed to be exchangeable in some sense with those already generated from the process under scrutiny. Clearly hypothesis testing and estimation as stressed in almost all statistics books involve parameters.

Hence this presumes the truth of the model and imparts an inappropriate existential meaning to an index or parameter. Model selection, contrariwise, is a preferable activity because it consists of searching for a single model (or a mixture of several) that is adequate for the prediction of observables even though it is unlikely to be the "true" one. This is particularly appropriate in those softer areas of application, which are legion, where the so-called true explanatory model is virtually so complex as to be unattainable, if it exists at all.

We then realize that inferring about observables is more pertinent since they can occur and be validated to a degree that is not possible with parameters. Divesting ourselves of parameters would then be our goal if we could model observables without recourse to parameters. Strictly speaking, in the completely observabilistic realm, modeling would require a finite number of observables whose measurements are discrete. The completely observabilistic view was brought particularly to the attention of British and American statisticians with the translation into English of the book on probability authored by de Finetti (1974). In this approach a well-defined and unique joint distribution for observables generated by some process is postulated. A predictive inference for unobserved values results from their distribution conditioned on those values already observed. This becomes, in most cases, such a difficult and unappealing enterprise that we tend to fall back on the infinite and the continuous and the parametric for ease in modeling. Indeed, certain parameters can be properly conceived as a result of passing to a limit of a function of observables. This need not overly concern us as long as we focus on the appropriate aspects of this endeavor. We must acknowledge that even aside from the measurement model, there are several other advantages to using a parametric model. Suppose our interest is in the prediction of the function $g(X_{N+1}, \ldots, X_{N+M})$ of future or unobserved values after a sample $X_1, \ldots, X_N$ is at hand. When M, the number of future values, is large so that exact computation is overwhelmingly burdensome, then going to the limit, i.e., as $M \to \infty$, may provide a simplification that is more than adequate for the purposes in mind. This limiting function, properly defined, can be considered a parameter and is a reasonable one in the context of the problem. In fact when no particular known M is at issue the conception of a hypothetically infinite number of as yet unobserved values may be useful as a normative evaluation procedure in making comparisons among therapeutic agents, treatments, etc. Since compelling physical models are often unavailable or too complex, as is usually the case in much statistical practice, we may be constrained to use statistical models. To then estimate or test certain of these parameters without embedding them in a limiting predictivistic interpretation vests them with an authority they do not possess. Hence, for statistical models, clearly the stress should be on observables and model selection cum prediction.

There is an important connection between statistical parametric modeling and the completely observable approach. If the observables are exchangeable, then the de Finetti representation theorem implies the existence of a "lurking parameter" or more generally the parametric model.

Most statistical analyses currently use tests of significance about parameters to form conclusions. A lesser number try to infer about parameters either as point or interval estimates. It would appear that if these "parameters" were known there would be no need for any analysis. The point of view expressed here is not whether a parameter is significantly different from some value or is in some interval but the effect of all this on the distribution of observables. In other words, the real analysis starts after we have made allowance for parameters, known or unknown. This point of view can resolve the often thorny issue of practical parametric significance versus statistical significance.

As regards statistical prediction, the amount of structure one can reasonably infuse into a given problem or process could very well determine the inferential model, whether it be frequentist, fiducialist, likelihood, or Bayesian. Any one of them possesses a capacity for implementing the predictive approach, but only the Bayesian mode is always capable of producing probability distributions for prediction.

For a number of examples of predictive inference that serve as a more sensible approach than parametric inference see Geisser (1971, 1988) and Aitchison and Dunsmore (1975).

One area that has always served as a natural enterprise for prediction or forecasting is time series. Books that stress time series forecasting such as Box and Jenkins (1970) have an orientation that is generally non-Bayesian. Most other non-Bayesian books in this area such as Anderson (1971) stress parametric estimation. Recently West and Harrison (1989) have delivered a Bayesian forecasting volume. As this material is given ample consideration there, we touch only on time series models. The main purpose of this book then is to propose and display methods for inferences or decisions that are more general and more informative about questions that statisticians often are asked. We say more general in that an inference about a parameter can also be framed as an inference about the limiting case of a function of future observables.

The chapters are organized so as to acquaint the reader first with non-Bayesian methods of prediction. This is the burden of Chapter 2. Since most statisticians and users are much more familiar with frequentist and other classical methods, this chapter should serve as an introduction to the techniques in changing the stress from estimation to prediction using familiar modes of inference. Once this is grasped they are invited to see the advantage of the Bayesian predictive approach displayed in the subsequent chapters. Convinced Bayesians may want to skim Chapter 2 to acquaint themselves with what non-Bayesians have proposed for prediction. However, they still may profit from Section 4 where only low structure modeling is available.

The rest of the chapters range over various topics where the predictive approach is appropriate. These methods involve, aside from straightforward predictions, comparisons, classifications, model selection, model adequacy, model perturbation, discordancy assessment, diagnostic screening, calibration and regulation, and interim analyses. Obviously the alert reader will easily identify many others. In fact it is opined that there are very few methods that cannot be put into a predictive framework.

References

Aitchison, J., and Dunsmore, I. R. (1975). *Statistical Prediction Analysis.* Cambridge: Cambridge University Press.

Anderson, T. W. (1971). *The Statistical Analysis of Time Series.* New York: Wiley.

Box, G. E. P., and Jenkins, M. (1970). *Time Series Analysis Forecasting and Control.* San Francisco: Holden-Day.

de Finetti B. (1974). *Theory of Probability*. New York: Wiley (first published in 1970 under title of *Teoria Delle Probabilita*).

Geisser, S. (1971). The inferential use of predictive distributions. In *Foundations of Statistical Inference*, V. P. Godambe and D. A. Sprott (eds.). Toronto: Holt, Rinchart & Winston, 456–469.

Geisser, S. (1988). The future of statistics in retrospect. In *Bayesian Statistics 3*, J. M. Bernardo et al. (eds.). Oxford: Oxford University Press, 147–158.

West, M., and Harrison, J. (1989). *Bayesian Forecasting and Dynamic Models*, Berlin: Springer.

CHAPTER 2

Non-Bayesian predictive approaches

In the previous chapter we delineated reasons for the importance of stressing observables. In this chapter we shall be concerned with various ways that have been devised to accommodate statistical predictions that are not Bayesian. These will include the classical confidence approach, various methods that depend on the likelihood, on loss functions, and on sample reuse procedures. The material here is largely for those who are inclined toward predictive analysis but are not entirely comfortable with a Bayesian approach. However, they are advised to continue with subsequent chapters to become acquainted with the large variety of problems for which the Bayesian approach provides solutions.

2.1 Confidence regions for future realizations

2.1.1 *Parametric case*

Let $X^{(N)} = (X_1, \ldots, X_N)$ be the set of values that will be observed in an experiment or series of trials and

$$X_{(M)} = (X_{N+1}, \ldots, X_{N+M})$$

be the unobserved set about which inference is to be made or action to be taken. We assume that the joint distribution of both sets is

$$F\left(x^{(N)}, x_{(M)} | \theta\right).$$

The frequentist confidence approach is to calculate the probability that the random vector $X_{(M)}$ falls in a random region which depends on $X^{(N)}$,

$$\Pr\left[X_{(M)} \in R_\alpha(X^{(N)}) | \theta\right] = 1 - \alpha.$$

If the region can be so formed that this probability does not depend on θ, then we interpret the above, once $X^{(N)} = x^{(N)}$ is realized, as the degree of confidence $1 - \alpha$ assigned to unobserved $X_{(M)}$ falling into $R_\alpha(x^{(N)})$. Such statements will be correct $100(1 - \alpha)\%$ in repeated sampling (in the long run) of $X^{(N)}$ and $X_{(M)}$ for a fixed $1 - \alpha$.

In some cases one would be required to predict a future scalar or vector-valued function $g(X_{(M)})$, noting that $X_{(M)}$ can be considered as a special case of $g(X_{(M)})$.

Easy implementation will generally require a scalar or vector valued pivotal function

$$P\left[g(X_{(M)}), h(X^{(N)})\right],$$

whose distribution is independent of θ, that can be conveniently inverted such that

$$\Pr\left[g(X_{(M)}) \in R_\alpha(h(X^{(N)}))\right] = 1 - \alpha.$$

where R_α is a region depending on $h(X^{(N)})$ the values to be observed. Such a pivot, containing only observables, is termed an ancillary and often requires a particular structure on the observational distribution involving the parameters and the inference to be made. Examples of this follow.

Example 2.1. Suppose $X_1, \ldots, X_N, X_{N+1}$ are independent copies from an $N(\theta, 1)$ distribution,

$$f(x | \theta) = \frac{1}{\sqrt{2\pi}} e^{-(1/2)(x - \theta)^2}, \qquad x \in R^1.$$

Then for $N\bar{X} = \Sigma_1^N X_i$,

$$\frac{X_{N+1} - \bar{X}}{\sqrt{1 + (1/N)}} \sim N(0,1)$$

and for $a < b$

$$\Pr\left[\bar{X} + a\frac{\sqrt{N+1}}{\sqrt{N}} \leq X_{N+1} \leq \bar{X} + b\frac{\sqrt{N+1}}{\sqrt{N}}\right] = \Phi(b) - \Phi(a)$$

$$= 1 - \alpha$$

where a and b depend on the fixed α and $\Phi(\cdot)$ is the standard normal distribution function,

$$\Phi(x) = \int_{-\infty}^{x} \frac{1}{\sqrt{2\pi}} e^{-(1/2)y^2} \, dy.$$

This implies that

$$\left(\bar{x} + a\frac{\sqrt{N+1}}{\sqrt{N}}, \qquad \bar{x} + b\frac{\sqrt{N+1}}{\sqrt{N}}\right)$$

is the predictive confidence interval for X_{N+1}. Note that the interval is random so it cannot be a fixed arbitrary interval chosen beforehand. This weakness persists for all applications of the predictive confidence approach. Setting $a = -b$ will minimize the length of the interval for a fixed probability and hence for a given confidence coefficient.

Note if we estimate the $N(\theta, 1)$ distribution function by its maximum likelihood estimator $N(\bar{x}, 1)$ and use this to estimate an interval for X_{N+1}, we would obtain a narrower interval since

$$\hat{\Pr}\left[\bar{X} + a \leq X_{N+1} \leq \bar{X} + b\right] = \Phi(b) - \Phi(a)$$

yields

$$\bar{x} + a \leq X_{N+1} \leq \bar{x} + b$$

but which cannot have the repeated sampling frequency interpretation for the assumed $1 - \alpha$.

Example 2.2. Suppose the variance is also unknown and $X_i \sim N(\mu, \sigma^2)$, $i = 1, \ldots, N + 1$. Then for

$$(N - 1)s^2 = \sum_1^N (X_i - \overline{X})^2,$$

$$\frac{X_{N+1} - \overline{X}}{s\sqrt{1 + (1/N)}} = t_{N-1} \sim S(t; N - 1)$$

where $S(t; \nu)$ is the "Student's t" distribution function with ν degrees of freedom with density

$$f(t) = \frac{\Gamma((\nu + 1)/2)}{[\pi\nu]^{1/2}\Gamma(\nu/2)}\left(1 + \frac{t^2}{\nu}\right)^{-(\nu+1)/2}, \qquad t \in R^1.$$

Hence for $b > a$, $\overline{X}$ and s random,

$$\Pr\left[\overline{X} + as\frac{\sqrt{N+1}}{\sqrt{N}} \le X_{N+1} \le \overline{X} + bs\frac{\sqrt{N+1}}{\sqrt{N}}\right]$$

$$= S(b) - S(a) = 1 - \alpha,$$

$$S(b; N - 1) - S(a; N - 1)$$

$$= \int_a^b f(t)\,dt.$$

Letting $a = -t_\alpha$ and $b = t_\alpha$ will yield the shortest confidence interval for a given confidence coefficient $1 - \alpha$. Note that the estimation approach that substitutes the maximum likelihood estimators would yield

$$\hat{\Pr}\left[\overline{X} - t_\alpha s\frac{\sqrt{N-1}}{\sqrt{N}} \le X_{N+1} \le \overline{X} + t_\alpha s\frac{\sqrt{N-1}}{\sqrt{N}}\right]$$

$$= \Phi(t_\alpha) - \Phi(-t_\alpha)$$

which could be substantially narrower than the one based on an exact frequency coverage. Obviously the latter could only serve as an approximation for sufficiently large N.

A further complication ensues when we consider not only the next observation but a whole set of them $X_{N+1}, \ldots, X_{N+M}$. Consider the random vector

$$Y' = (Y_1, \ldots, Y_M)$$

where

$$Y_i = X_{N+i} - \overline{X}, \qquad i = 1, \ldots, M$$

so that Y is multivariate normal $N(0, \sigma^2 \Omega)$,

$$f(y) \propto e^{-(1/2\sigma^2)y'\Omega^{-1}y}, \qquad y \in R^M$$

for

$$\Omega = I + N^{-1}ee',$$
$$\Omega^{-1} = I - (N+M)^{-1}ee'$$

where e is an M-dimensional vector all of whose components are one. Since

$$(N-1)s^2 \sim \sigma^2 \chi^2_{N-1}$$

independently of Y, the density of the vector

$$Z = \frac{Y}{\sqrt{(N-1)s^2}}$$

is easily shown to be a multivariate student density

$$f(z) = \frac{\Gamma((N+M-1)/2)}{\pi^{M/2}\Gamma((N-1)/2)(N-1)^{M/2}}[1 + z'\Omega^{-1}z]^{-(N+M-1)/2},$$

$$z \in R^M$$

denoted as $Z \sim S_M(N - 1, 0, \Omega)$. In general $Z \sim S_M(\nu, \mu, \Sigma)$ if

$$f(t) = \frac{\Gamma((\nu + M)/2)|\Sigma|^{-1/2}}{\pi^{M/2}\Gamma(\nu/2)}\left[1 + (z - \mu)'\Sigma^{-1}(z - \mu)\right]^{-(\nu + M)/2}$$

From the above we can easily show that

$$\frac{N-1}{M}Z'\Omega^{-1}Z \sim F_{M, N-1}$$

an F distribution with M and $N - 1$ degrees of freedom, where $X \sim F_{a,b}$ has density

$$f(x) \propto x^{a/2-1}\left(1 + \frac{a}{b}x\right)^{-(a+b)/2}, \qquad 0 < x, a \geq 1, b \geq 1.$$

Hence a confidence hyperellipse is readily obtainable from

$$\Pr\left[Z'\Omega^{-1}Z \leq \frac{M}{N-1}F_\alpha\right] = 1 - \alpha,$$

yielding all values for $X_{N+1}, \ldots, X_{N+M}$ that lie in the hyperellipse centered at $(\bar{x}, \ldots, \bar{x})$ namely

$$\left(X_{N+1} - \bar{x}, \ldots, X_{N+M} - \bar{x}\right)\Omega^{-1}\begin{pmatrix} X_{N+1} - \bar{x} \\ \vdots \\ X_{N+M} - \bar{x} \end{pmatrix} \leq Ms^2 F_\alpha.$$

Numerical illustration of Example 2.2. In a random sample from a normal distribution, the following values were obtained: -0.109, 0.404, -0.920, -1.208, -0.499, 0.064, and 0.629. For $M = 2$, and arbitrary sample size N,

$$\Omega^{-1} = \begin{pmatrix} 1 - \dfrac{1}{N+2} & -\dfrac{1}{N+2} \\ -\dfrac{1}{N+2} & 1 - \dfrac{1}{N+2} \end{pmatrix}$$

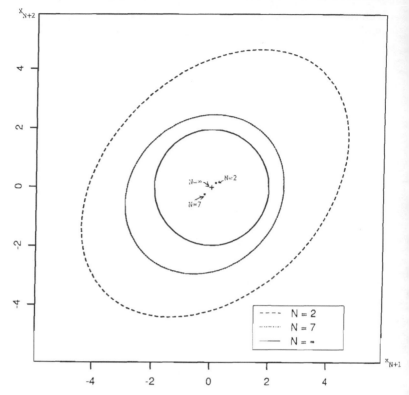

Figure 2.1   95% contours of $f(x_{N+1}, x_{N+2}|x^{(N)})$ with centers $N = 2$, $N = 7$ and $N = \infty$

To observe the effect of increasing N, Figure 2.1 displays the predictive confidence ellipsis for three cases:

i. $N = 2$, only the first two observations are used
ii. $N = 7$, the entire sample is used
iii. $N \to \infty$ (the observations were actually a sample from an $N(0, 1)$ distribution).

Note how the ellipses tend asymptotically to a circle (independence for $N = \infty$).

Suppose now we were interested in the number of these M future observations that will lie in a particular interval. Problems of this sort can arise when, for example, a customer is ordering M

items and quality requires these items to be within a given interval. Within this framework we can obtain an interval only of the following kind,

$$J_a = \bar{X} \pm as,$$

where a depends on α and N. Let

$$\Pr[X_{N+i} \in J_a] = \Pr[Z_i \in I_a]$$

where $I_a = (-a, a)$. Let

$$V_i = 1 \qquad \text{if } Z_i \in I_a$$
$$V_i = 0 \qquad \text{otherwise}$$

and $R = \sum_{i=1}^{M} V_i$. Then

$$\Pr[R = r] = \binom{M}{r} \Pr[Z_1 \in I_a, \dots Z_r \in I_a, Z_{r+1} \in I_a^c, \dots Z_M \in I_a^c]$$

can be obtained from the multivariate student density of Z. This may be quite burdensome for moderate or large M. There is some simplification if M is sufficiently large and the interval is

$$I = (-\infty, \bar{X} + as).$$

By sufficiently large we mean that M is large enough so that by defining

$$V = \lim_{M \to \infty} \frac{R}{M}$$

that

$$P(V \geq v) \doteq \Pr\left(\frac{R}{M} \geq v\right).$$

Further V may itself be of interest as the fraction of all future $X_{N+1}, X_{N+2}, \dots$ that lie in I. Clearly then

$$V = \Pr[X \leq \bar{X} + sa \,|\, \sigma, \mu] = \Phi\left(\frac{\bar{X} - \mu}{\sigma} + \frac{as}{\sigma}\right).$$

Since

$$\frac{\bar{X} - \mu}{\sigma} \sim N\left(0, \frac{1}{N}\right)$$

$$W = \frac{s}{\sigma} \sim \sqrt{\chi^2_{N-1}/(N-1)} \,,$$

it follows that

$$\Pr[V \geq v] = \Pr\left[\Phi\left(\frac{1}{\sqrt{N}} Y + aW\right) \geq v\right]$$

where $Y \sim N(0,1)$ and independent of W. Hence

$$\Pr[V \geq v] = \Pr\left[\frac{1}{\sqrt{N}} Y + aW \geq \Phi^{-1}(v)\right]$$

$$= \Pr\left[\frac{\sqrt{N}\, \Phi^{-1}(v) - Y}{W} \leq \sqrt{N}a\right].$$

The random variable in the bracket to the left of the inequality, say T, is clearly distributed as a noncentral "student" variable, with $\nu = N - 1$ degrees of freedom and noncentrality parameter $\gamma = \Phi^{-1}(v)\sqrt{N}$ independent of μ and σ^2. This is denoted as $T \sim S(t; \nu, \gamma)$. The density of T is given as

$$f(t) = \frac{e^{-(1/2)\gamma^2}}{\sqrt{\pi \nu}\, \Gamma(\nu/2)} \sum_{j=0}^{\infty} \left(1 + \frac{t^2}{\nu}\right)^{-(\nu+1+j)/2} \frac{\Gamma((\nu + 1 + j)/2)(t\gamma\sqrt{2})^j}{j!\nu^{j/2}}$$

Hence we may set

$$\sqrt{N}a = t_\alpha\left[N - 1, \sqrt{N}\, \Phi^{-1}(v)\right]$$

where the quantity on the right is the appropriate percentage point of the noncentral t-distribution and solve for a fixed v and α to obtain confidence limits on V.

In all of these examples we have chosen particular pivots out of the many that could have been chosen. In Example 2.2 another possible pivot is

$$Z = \frac{(X_{N+1} - \bar{X})\sqrt{N}}{U\sqrt{N+1}}$$

where

$$U^2 = \frac{1}{2(N-1)} \sum_{i=1}^{N-1} (X_{i+1} - X_i)^2$$

is the mean square successive difference. It can be shown that the numerator and denominator of Z are independent (Geisser, 1956), and that Z is an ancillary. Hence a predictive confidence interval for X_{N+1} is available. The point then would be to discriminate among pivots to obtain an optimum one for some purpose. A theory of optimal predictive pivots may be addressed in much the same way as a theory of optimal parametric pivots via loss functions.

2.1.2 Low structure stochastic case

Suppose $X_1, X_2, \ldots, X_N, X_{N+1}, \ldots, X_{N+M}$ are absolutely continuous and exchangeable, so that every permutation is equally likely. Let $Y_1 < Y_2 < \cdots < Y_N$ be the order statistics for the sample $X_1, \ldots, X_N$. Further define $Y_0 = -\infty$ and $Y_{N+1} = \infty$, then it is clear that

$$\Pr\left[Y_j < X_{N+1} < Y_{j+1}\right] = \frac{1}{N+1}, \qquad j = 0, \ldots, N$$

and

$$\Pr\left[Y_j < X_{N+1} < Y_{j+k}\right] = \frac{k}{N+1}, \qquad k = 1, \ldots, N-j+1.$$

This then is the probability that a random variable lies in a random interval. Hence with confidence coefficient $1 - \alpha = k/(N+1)$, we obtain the confidence interval for the future X_{N+1} as (y_j, y_{j+k}). Note the restriction on the type of intervals and confidence coefficients that can be chosen. More generally we can calculate the probability that R, the random number of M future

Xs, is in the random interval (Y_j, Y_{j+k}). Consider all sequences such that exactly u of the Xs are in (Y_0, Y_j), r in (Y_j, Y_{j+k}) and $M - r - u$ in (Y_{j+k}, Y_{N+1}). Summing over u from 0 to $M - r$ we obtain the number of ways of having exactly r in (Y_j, Y_{j+k}). This is then divided by the total number of ways of distributing the M future Xs among the Ys. Hence

$$\Pr(R = r)$$

$$= \frac{\binom{r + k - 1}{r}}{\binom{M + N}{N}} \sum_{u=0}^{M-r} \binom{u + j - 1}{u} \bigg/ \binom{N - j - k + M - r - u}{M - r - u}.$$

The quantity to the right of the summation sign is the number of ways of drawing u red balls from $M - r$ red balls given that sequential sampling stops when we have drawn j out of $N - k$ white balls from the urn containing the red and white balls. Hence that quantity is recognized as a negative hypergeometric probability function when divided by

$$\binom{N - k + M - r}{N - k}.$$

Thus the sum is equal to the above and we obtain

$$\Pr[R = r] = \frac{\binom{r + k - 1}{r}\binom{N + M - r - k}{M - r}}{\binom{N + M}{N}} = 1 - \alpha$$

which is also recognizable as a negative hypergeometric probability function. Then with confidence coefficient $1 - \alpha$, R of the M future Xs lie in (y_j, y_{j+k}). As a numerical example let $N = 15$, $M = 5$, and $k = 5$, so that R is the number of the M future Xs in (y_j, y_{j+5}). Notice that the results in Table 2.1 are valid for any integer $j = 1, \ldots, 10$.

Again, if we let

$$V = \lim_{M \to \infty} \frac{R}{M}$$

Table 2.1 *Calculation of* $\Pr(R = r) = 1 - \alpha$

r	0	1	2	3	4	5
$P(R = r)$	0.1937	0.3228	0.2767	0.1490	0.0497	0.0081

be the fraction of all future observations that lie in the random interval (Y_j, Y_{j+k}), it is easy to show that

$$V \sim \beta(v; k, N - k + 1)$$

a β variate with density

$$f(v) = \frac{N!}{(k - 1)!(N - k)!} v^{k-1}(1 - v)^{N-k}.$$

Clearly then we also have that

$$V = F_X(Y_{j+k}) - F_X(Y_j)$$

where $F_X(\cdot)$ is the distribution function of X. Now for $0 \le \gamma < \gamma' \le 1$,

$$\Pr[\gamma < V < \gamma'] = \alpha' - \alpha = \int_{\gamma}^{\gamma'} f(v)\, dv,$$

so that $\alpha' - \alpha$ is the confidence coefficient that the proportion of all future Xs that lie in (y_j, y_{j+k}) is between γ and γ'. If we set $\gamma' = \alpha' = 1$, then with confidence $1 - \alpha$, the percent of the population in (y_j, y_{j+k}) is at least $100\gamma\%$.

2.2 Estimating a parametric predicting density

We have already noticed that introducing maximum likelihood estimates for the mean and variance into a normal distribution results in estimated prediction intervals that are too tight in the frequency sense. We now examine another method of maximizing to obtain an estimated predicting density. Such a density and its distribution function are useful in obtaining estimates of the probability that a future observation will fall in some prescribed

interval region or set. It can also be used to identify probabilisti-
cally one of several populations from which a new observation
could have arisen.

2.2.1 Maximum likelihood predicting density (MLPD)

A maximum likelihood approach for estimating a predicting density was devised by Lejeune and Faulkenberry (1982). For inde-
pendent X_j, $j = 1, \ldots, N + M$, with density $f(x|\theta)$, define

$$\hat{f}\left(x_{(M)}|x^{(N)}\right) \propto \sup_{\theta} f\left(x^{(N)}|\theta\right) f\left(x_{(M)}|\theta\right) \qquad (2.1)$$

or for a function $g(x_{(M)}) = g$

$$\hat{f}\left(g|x^{(N)}\right) \propto \sup_{\theta} f\left(x^{(N)}|\theta\right) f(g|\theta). \qquad (2.2)$$

This approach will correspond only rarely to a frequentist approach (Example 2.4). Under certain stringent conditions it will
correspond to a Bayesian approach as in Example 2.3.

Example 2.3. Let X_j, $j = 1, \ldots, N + 1$ be independently distributed as $N(\mu, \sigma^2)$. Then for $\theta = (\mu, \sigma^2)$

$$\hat{f}\left(x_{N+1}|x^{(N)}\right) \propto \sup_{\theta} \frac{1}{\sigma^{N+1}} e^{-(1/2\sigma^2) \sum_{1}^{N+1} (x_i - \mu)^2}$$

$$\propto \left(1 + \frac{(x_{N+1} - \bar{x})^2}{s^2(1 + (1/N))(N - 1)}\right)^{-(N+1)/2}$$

since

$$\sum_{1}^{N+1} (x_i - \bar{x}_{N+1})^2 = (N - 1)s^2 + \frac{N}{N + 1}(x_{N+1} - \bar{x})^2.$$

Hence

$$\frac{X_{N+1} - \bar{x}}{s(1 + (1/N))^{1/2}} \sim \left(1 - \frac{1}{N}\right)^{1/2} t_N.$$

But the quantity on the left has a frequency density of t_{N-1} so that this result will not have the appropriate frequency interpretation. However, it can be obtained via a Bayesian approach for σ and μ both uniformly distributed a priori. Further

$$\lim_{N \to \infty} f\left(x_{N+1}|x^{(N)}\right) = \frac{e^{-(1/2\sigma^2)(x_{N+1}-\mu)^2}}{\sigma\sqrt{2\pi}},$$

so that it converges appropriately.

Example 2.4. Let X_i, $i = 1, \ldots, M + N$ be independent with density

$$f(x|\theta) = \theta e^{-\theta x},$$

then

$$\hat{f}\left(x_{(M)}|x^{(N)}\right) \propto \sup_{\theta} \theta^N e^{-\theta \sum_1^N x_i} \theta^M e^{-\theta \sum_{i=1}^M x_{N+i}}.$$

Since

$$\hat{\theta} = (N + M) \Big/ \sum_1^{N+M} x_i,$$

$X_{(M)}$ is assigned density

$$\hat{f}\left(x_{(M)}|x^{(N)}\right) \propto \left(N\bar{x} + x_{N+1} + \cdots + x_{N \cdot M}\right)^{-(N+M)}$$

where $N\bar{x} = \sum_1^N x_i$. Note here that for fixed M

$$\lim_{N \to \infty} \hat{f}\left(x_{(M)}|x^{(N)}\right) = \theta^M \exp\left[-\theta \sum_{i=1}^M x_{N+i}\right]$$

the original sampling density of $X_{(M)}$. Note also that the assigned density of vector $U = X_{(M)}/\sum_1^N x_i$ corresponds to the appropriate sampling density of $U = X_{(M)}/\sum_1^N X_i$ as can readily be verified.

Example 2.5. Let X_i, $i = 1, \ldots, N + M$ with

$$\Pr[X_i = 1] = \theta = 1 - \Pr[X_i = 0].$$

Let $S = \sum_1^N X_i$ and $R = \sum_1^M X_{N+i}$ then

$$\hat{f}(r|s) \propto \sup_\theta \binom{N}{s} \theta^s (1-\theta)^{N-s} \binom{M}{r} \theta^r (1-\theta)^{M-r}.$$

Hence

$$\hat{\theta} = \frac{r+s}{N+M}$$

and

$$\hat{f}(r|s) = \frac{\binom{M}{r}(r+s)^{r+s}(N+M-r-s)^{N+M-r-s}}{\sum\limits_{j=0}^{M} \binom{M}{j}(j+s)^{j+s}(N+M-j-s)^{N+M-j-s}}.$$

For $M = 1$

$$\lim_{N \to \infty} \hat{f}(r|s) = \begin{cases} \theta & \text{if } r = 1 \\ 1 - \theta & \text{if } r = 0 \end{cases}$$

i.e., the sampling probability of X_{N+1}.

For sufficiently large N

$$\hat{f}(r|s) \doteq \binom{r+s-1/2}{r}\binom{N+M-r-s-1/2}{M-r} \Big/ \binom{N+M}{N}$$

is obtained by using the factorial notation for gamma functions $a! = \Gamma(a+1)$, $a \geq 0$, Stirling's formula,

$$a! \sim \sqrt{2\pi} \, / (a+1)\left(\frac{a+1}{e}\right)^{a+1},$$

and noting that as a grows

$$\frac{a!}{(a - (1/2))!\sqrt{a}} \to 1.$$

Lejeune and Faulkenberry also provide a theorem that indicates when their procedure is equivalent to a Bayesian approach. Leonard (1982) provides some approximations of the MLPD for large samples. He applies Bayes theorem in reverse using a large sample approximation for the posterior density of θ given both $x^{(N)}$ and $x_{(M)}$ to obtain an approximate MLPD.

2.2.2 Loss function estimators of predicting densities

Define

$$L(f, g)$$

where $f = f(x_{(M)}|\theta)$ and $g = g(x_{(M)}|x^{(N)})$ to be the loss in using g to estimate f when θ is the true value. Let

$$\bar{L}(f, g) = E_{X_{(M)}}[L(f, g)]$$

where the expectation is over f. Now for $g \in G$, G being a defined class of densities, let

$$E_{X^{(N)}}\bar{L}(f, g) = \bar{\bar{L}}(f, g)$$

and find that $g = \hat{f}$ such that

$$\inf_{g \in G} \bar{\bar{L}}(f, g) = \bar{\bar{L}}(f, \hat{f}).$$

If $\hat{f}$ exists and is unique and does not depend on θ, then $\hat{f}$ is the optimal predicting density estimator in terms of minimizing expected loss. A class of such loss functions is the Hellinger distance,

$$\bar{L} = H(f, g) = E\left|1 - \left(\frac{g}{f}\right)^{1/n}\right|^n.$$

However it is, in general, analytically awkward for $n \neq 2$. A more amenable loss function is the Kullback and Leibler (1951) directed divergence

$$\overline{L} = K(f, g) = E\left[\log \frac{f}{g}\right].$$

This has been proposed and used for particular problems by Aitchison (1975) and Murray (1977) with certain constraints on g. However, the existence of a unique solution independent of θ appears to be restricted to certain families of distributions such as the normal and gamma.

Example 2.6. Let X_i $i = 1, \ldots, N + M$ be independently distributed with density

$$f(x) = \theta e^{-\theta x}.$$

Knowing that the vector

$$\left(\frac{X_{N+1}}{N\overline{X}}, \ldots, \frac{X_{N+M}}{N\overline{X}}\right) = \frac{X_{(M)}}{N\overline{X}} = U' = (U_1, \ldots, U_M)$$

is invariant under changes of scale where $N\overline{X} = \sum_1^N X_i$ we shall restrict attention to estimates g that are invariant under changes of scale or

$$h_U(u) = (N\overline{x})^M g_{X_{(M)}}\left(\frac{X_{(M)}}{N\overline{x}}\right).$$

Then

$$K(f, g) = E \log \frac{f\left(X_{(M)}|\theta\right)}{g\left(\dfrac{X_{(M)}}{N\overline{X}}\right)}$$

where

$$f\left(x_{(M)}|\theta\right) = \theta^M e^{-\theta \sum_1^M x_{N+i}}$$

and

$$K(f, g) = E\left[M \log \theta - \theta \sum_{1}^{M} X_{N+i} - \log g\left(\frac{X_{(M)}}{N\bar{x}} \right) \right]$$

$$= M \log \theta - M - E \log g\left(\frac{X_{(M)}}{N\bar{x}} \right),$$

where the expectation is over $X_{(M)}$.
    Let

$$I(f, g) = E_{X^{(N)}}[K(f, g)]$$

then minimizing $I(f, g)$ is equivalent to maximizing

$$E_{X^{(N)}} E_{X_{(M)}}\left[\log g\left(\frac{X_{(M)}}{N\bar{X}} \right) \right] = \int f_U(u|\theta) \log g_U(u)\, du$$

where the vector $U = X_{(M)}/N\bar{X}$ is clearly distributed independently of θ and has density, say $f_U(u)$. Hence we need to maximize

$$\int f_U(u) \log g_U(u)\, du,$$

where

$$g_U(u) = (N\bar{X})^M g_{X_{(M)}}(u).$$

    We cite a standard result in information theory Rao (1965, p. 47)

$$\int f \log f\, du \geq \int f \log g\, du$$

with equality if and only if $f = g$ a.e.

Hence the integral is maximized for $g_U(u) = f_U(u)$. Now it is easily shown that the sampling density of U is

$$f_U(u) = \frac{\Gamma(N+M)}{\Gamma(N)(1 + u_1 + \cdots + u_M)^{N+M}}$$

hence

$$\hat{f}_U(u) = f_U(u)$$

and

$$\hat{f}_{X_{(M)}}\left(x_{(M)}|x^{(N)}\right) = \frac{\Gamma(N+M)(N\bar{x})^N}{\Gamma(N)(N\bar{x} + x_{N+1} + \cdots + x_{N+M})^{N+M}}.$$

Example 2.7. Let X_i, $i = 1, \ldots, N + M$ be independently distributed as $N(\mu, \sigma^2)$ where g is restricted to be of the form

$$g\left(\frac{\left(X_{(M)} - \bar{X}e'\right)}{s\sqrt{1 + (1/N)}}\right), \qquad e' = (1, \ldots, 1).$$

Note that the sampling distribution of the components of g does not depend on μ or σ.

Similar to the previous example we can show that minimizing $I(f, g)$ yields

$$\hat{f}\left(x_{(M)}|x^{(N)}\right) \propto \left(1 + \frac{1}{(N-1)s^2}\left(x'_{(M)} - \bar{x}e\right)'\right.$$

$$\left. \times \Omega^{-1}\left(x'_{(M)} - \bar{x}e\right)\right)^{-(N+M-1)/2}$$

a multivariate student density for $X_{(M)}$, where $\Omega = I + N^{-1}ee'$,

$$S_M\left[N - 1, \bar{x}e, (N-1)s^2\Omega\right].$$

If

$$Z' = \frac{X_{(M)} - \overline{X}e'}{s\sqrt{N-1}}$$

then

$$\hat{f}(z) \propto (1 + z'\Omega^{-1}z)^{-(N+M-1)/2}$$

and

$$\frac{N-1}{M}Z'\Omega^{-1}Z \sim F_{M, N-1}.$$

Thus agreement with existing predictive confidence regions with frequentist interpretation is obtained. However, in this example as well as in the previous one the estimated predictive distributions may be used in a variety of ways that do not concord with frequentist confidence statements. Further, this method is useful only when the predicting density is optimal for all values of θ, which indicates that the condition placed on g is equivalent to using a pivot. Thus the procedure is quite restricted.

2.3 Methods based on likelihoods

2.3.1 *Fisher relative likelihood (second-order likelihood)*

There are methods that according to their adherents provide a scaling of the likelihood of future values based on observed values. Let the relative likelihood be defined as

$$R(\theta, x) \equiv \frac{f(x|\theta)}{f(x|\hat{\theta})} = \frac{L(\theta)}{L(\hat{\theta})}$$

where $L(\theta)$ is the likelihood of θ and finite for all θ,

$$f(x|\hat{\theta}) = \sup_{\theta} f(x|\theta).$$

Then Fisher (1956) defines the relative likelihood for independent and identically distributed values, when it exists, as

$$\text{FRL}\left(x_{(M)}|x^{(N)}\right) = \sup_{\theta} R(\theta, x^{(N)}) R(\theta, x_{(M)})$$

which is used to order the plausible values for $x_{(M)}$. He asserts that this procedure is to be used only in cases where another argument (fiducial) is unavailable. This would be the case for all discrete sampling distributions.

Example 2.8. Let X_i, $i = 1, \ldots, N + M$ be independent such that

$$P(X_i = 1|\theta) = \theta = 1 - \text{Pr}(X_i = 0|\theta).$$

The problem then is to predict the number of successes R out of M future trials assuming $S = s$ successes in the observed N trials. Clearly

$$L_N(\theta) = \theta^s(1 - \theta)^{N-s},$$

$$L_M(\theta) = \theta^r(1 - \theta)^{M-r}.$$

Then for $r = 0, \ldots, M$

$$\text{FRL}(r|s) \; \alpha \; \frac{(r + s)^{r+s}(M + N - r - s)^{N+M-r-s}}{r^r(M - r)^{M-r}}$$

purports to scale the likelihood of future values of r given s. The most "likely" value(s) is(are) achieved by maximizing FRL($r|s$).

This example was given by Fisher and the procedure detailed in the more general form by Kalbfleisch (1971). Its restricted use is exemplified by applying the method to the simple exponential case for $M = 1$.

Example 2.9. Let X_i, $i = 1, \ldots, N + 1$ be independently distributed with density

$$f(x|\theta) = \theta e^{-x\theta}$$

then

$$\text{FRL}\left(x_{N+1}|x^{(N)}\right) = \frac{x_{N+1}\bar{x}^N(N+1)^{N+1}}{\left(x_{N+1}+N\bar{x}\right)^{N+1}} \rightarrow \theta x_{N+1}e^{-\theta x_{N+1}}$$

as N grows, which provides a scaling for x_{N+1}, which differs markedly from the scaling for x_{N+1} provided by its density. Fisher does not recommend it as a fiducial argument is available for this case.

2.3.2 Predictive likelihood

This method for ordering the plausibility of future values was developed by Lauritzen (1974), Hinkley (1979), and somewhat more generally by Butler (1986). It involves transforming

$$\left(X^{(N)}, X_{(M)}\right) \rightarrow \left[T\left(X^{(N)}, X_{(M)}\right), U\left(X^{(N)}, X_{(M)}\right)\right]$$

where T is a minimal sufficient statistic and U denotes the remaining set of orthonormal coordinates such that

$$\frac{\partial(t,u)}{\partial\left(x^{(N)}, x_{(M)}\right)} = \left|\begin{matrix} J \\ K \end{matrix}\right| = \left|\begin{pmatrix} J \\ K \end{pmatrix}(J', K')\right|^{1/2} = \left|\begin{matrix} JJ' & 0 \\ 0 & I \end{matrix}\right|^{1/2} = |JJ'|^{1/2}$$

where

$$J = \frac{\partial t}{\partial\left(x^{(N)}, x_{(M)}\right)}, \qquad K = \frac{\partial u}{\partial\left(x^{(N)}, x_{(M)}\right)}, \qquad KJ' = 0, \; KK' = I$$

are the matrices of the indicated partial derivatives. Hence

$$dx^{(N)} dx_{(M)} = |JJ'|^{-1/2} du \, dt$$

and

$$f\left(x^{(N)}, x_{(M)}|\theta\right) dx^{(N)} dx_{(M)} = \frac{f(t,u|\theta)}{f(t|\theta)} |JJ'|^{-1/2} du f(t|\theta) \, dt.$$

Because of the sufficiency property of t, the first density on the right namely $f(u|t)$ is independent of θ. Hence Butler defines

$$PL_B\left(x_{(M)}|x^{(N)}\right) = \frac{f\left(x^{(N)}, x_{(M)}|\theta\right)}{f(t|\theta)|JJ'|^{1/2}}$$

to be the (conditional) predictive likelihood of $x_{(M)}$ given $x^{(N)}$ and uses it to order the plausibility of the values for $x_{(M)}$.

In cases where $X^{(N)}$ and $X_{(M)}$ are independent given θ

$$PL_B\left(x_{(M)}|x^{(N)}\right) = \frac{f(x^{(N)}|\theta)f\left(x_{(M)}|\theta\right)}{f(t_{N+M}|\theta)|JJ'|^{1/2}}$$

where $t_N = t(x^{(N)})$ is minimally sufficient based only on $x^{(N)}$ and $t_{N+M} = t(x^{(N)}, x_{(M)})$ is minimally sufficient based on $x^{(N)}$ and $x_{(M)}$. In the independence case Hinkley advocates

$$PL_H = f\left(x_{(M)}|t_M\right)f(t_M|t_{N+M}) = \frac{f\left(x_{(M)}|\theta\right)f(x^{(N)}|\theta)}{f\left(x^{(N)}|t_N\right)f(t_{N+M}|\theta)}$$

or

$$PL_H = |JJ'|^{1/2} PL_B\left(x_{(M)}|x^{(N)}\right)$$

since $f(x^{(N)}|t_N)$ does not contain $x_{(M)}$. Note also that the alternative form can be given as

$$PL_H = f\left(x_{(M)}|t_M\right)\frac{f(t_M|\theta)f(t_N|\theta)}{f(t_{N+M}|\theta)}.$$

When $(X^{(N)}, X_{(M)})$ are discrete and independent the two methods are the same, i.e.,

$$PL_H = PL_B = PL.$$

Example 2.10. Let X_i be independently distributed as

$$\Pr[X_i = 1|\theta] = \theta = 1 - \Pr[X_i = 0|\theta].$$

Then $S = \Sigma_1^N X_i$, $R = \Sigma_1^M X_{N+i}$, and

$$
PL(r|s) = \frac{\binom{N}{s}\theta^s(1-\theta)^{N-s}\binom{M}{r}\theta^r(1-\theta)^{M-r}}{\binom{N+M}{r+s}\theta^{r+s}(1-\theta)^{N+M-r-s}} = \frac{\binom{N}{s}\binom{M}{r}}{\binom{N+M}{r+s}}
$$

since $f(x_{(M)}|t_M)$ is constant.

Example 2.11. Let X_i be independently distributed as

$$
f(x|\theta) = \theta e^{-\theta x}
$$

then, since $t = \Sigma_1^{N+M} x_i$, $t_N = \Sigma_1^N x_i$, $t_M = \Sigma_1^M x_{N+i}$, $|JJ'| \propto$ Const., and

$$
PL_H\left(x_{(M)}|x^{(N)}\right)
$$

$$
= PL_B\left(x_{(M)}|x^{(N)}\right)
$$

$$
= \frac{(\theta^N t_N^{N-1} e^{-\theta t_N})/\Gamma(N)(\theta^M t_M^{M-1} e^{-\theta t_M})/\Gamma(M)}{\theta^{N+M}(t_N + t_M)^{N+M-1} e^{-\theta(t_N+t_M)}/\Gamma(N+M)} f\left(x_{(M)}|t_M\right).
$$

Hence omitting constants, the predictive likelihood is

$$
\frac{t_N^{N-1}}{(t_N + x_{N+1} + \cdots + x_{N+M})^{N+M-1}}.
$$

This result scales differently than the MLPD or the Kullback loss function predicting density approach but is asymptotically equivalent.

Numerical illustration of Examples 2.4, 2.6, 2.9, and 2.11. A set of eight observations was generated from an exponential distribution with mean 1. The observations were 4.038, 0.094, 2.744, 0.582, 0.317, 0.031, 0.193, 0.904. All predicting methods were then plotted on the same graph and all normed so that the area under each curve was unity to make the comparison easier to see (Fig. 2.2). Two of the methods MLPD and the K–L loss function approach yield the same result. The MLPD/KL result is quite close to the PL method. As expected, the extended FRL exhibits a completely different scaling of the predicted value of a future observable and would yield rather anomalous results, if used.

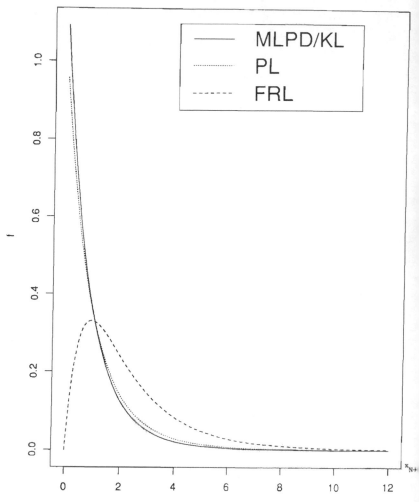

Figure 2.2   *Comparison of predicting densities and FRL from an exponential distribution.*

Example 2.12. For $X_i \sim N(\mu, \sigma^2)$ it can easily be shown that for $N \geq 3$

$$\mathrm{PL}_\mathrm{H}\left(x_{N+1}|x^{(N)}\right) = \left[1 + \frac{N(x_{N+1} - \bar{x})^2}{s^2(N^2 - 1)}\right]^{-(N-2)/2}$$

and

$$PL_B(x_{N+1}|x^{(N)}) = \frac{PL_H(x_{N+1}|x^{(N)})}{\left((N-1)s^2 + [N/(N+1)](x_{N+1} - \bar{x})^2\right)^{1/2}}$$

the former scaling x_{N+1} as a t with $N-3$ and the latter as a t with $N-2$ degrees of freedom, respectively. Again this is not equivalent to the MLPD or the Kullback loss function approach. The former scales as a t with N degrees of freedom and the latter as a t with $N-1$ degrees of freedom. However, all of these "densities" can be obtained by a Bayesian approach with prior density for μ and σ given by

$$p(\mu, \sigma) \propto \sigma^{-a}$$

where $a = 0, 1, 2, 3$ depending on which t distribution results.

2.4 Low structure nonstochastic prediction

2.4.1 Validation of predictors or models

Suppose a set $x^{(N)}$ of observations is randomly divided into two parts $x^{(N-n)} = (x_1, \ldots, x_{N-n})$ called the construction sample and $x^{(n)} = (x_{N-n+1}, \ldots, x_N)$ the validation sample. Assume that with each observation x_j there is associated a known value (or set of values) z_j and that given z_j the model is such that the x_js can be made fungible (exchangeable, if stochastic), e.g., $x_j = \alpha + \beta z_j + e_j$ so that $e_j = x_j - \alpha - \beta z_j$ are. Suppose further the relative values of K competing models for the data are to be assessed by comparing the predictions they make. Say that each model leads to a predictor

$$\hat{x}_{jk} = \hat{x}_{jk}\left(x^{(N-n)}, z^{(N-n)}; z_j\right)$$
$$j = N-n+1, \ldots, N \quad \text{and} \quad k = 1, \ldots, K.$$

A discrepancy $d_{jk} = d(\hat{x}_{jk}, x_j)$ can be computed for each model and the histogram or the empirical distribution function plotted and compared, e.g., $\hat{x}_{jk} - x_j = r_{jk}$. When the validation portion is not large enough to support such a comparison, one can compute relevant summary measures such as

$$s_k^2 = n^{-1} \sum_{j=N-n+1}^{N} r_{jk}^2 \quad k = 1, \ldots, K,$$

$$S_k = n^{-1} \sum_{j=N-n+1}^{N} |r_{jk}|,$$

or some scoring function such as the number of times $\hat{x}_{jk}$ is closer to x_j for model M_k as opposed to $M_{k'}$.

2.4.2 *Cross-validation*

Divide the sample in half, use the second half to "validate" the first half and vice versa, yielding a second validation or comparison. The two may be combined into a single one.

2.4.3 *Simple cross-validatory choice or predictive sample reuse*

Let $x_j^{(N-1)} = (x_1, \ldots, x_{j-1}, x_{j+1}, \ldots, x_N)$ and similarly for $z_j^{(N-1)}$. Let the kth predictive function for x_j be computed as

$$\hat{x}_{jk} = \hat{x}_{jk}\left(x_j^{(N-1)}, z_j^{(N-1)}, z_j\right)$$

and repeat this for $j = 1, \ldots, N$. Then compute, say, for the kth predictor $r_{jk} = \hat{x}_{jk} - x_j$ $j = 1, \ldots, N$ and compare the r_{jk} for the K models. Note that in the validation case the residuals are functionally dependent by virtue of the use of the same predictive function while in this case further functional dependence is infused because of using the data, repetitively. On the other hand this procedure uses all of the data in a symmetric fashion.

2.4.4 *Predictive sample reuse (general)*

Interest focuses on the prediction of a future observation or a set of such observations.

Ingredients and steps of the procedure

1. A predictive function for a future value x or set of values at given z is prescribed as

$$x(x^{(N)}, z^{(N)}, z; \omega) \qquad \omega \in \Omega$$

where ω is a set of unknown values either dictated by a model or some empirical considerations.

2. Let $P_i^{(N-n)}$ represent the ith partition of the original sample into $N - n$ retained observations and n omitted observations for $0 \leq n \leq M$ where M is the largest number such that the predictive function can be formed with $N - M$ observations. The observation set $x^{(N)}$ and its concomitant set $z^{(N)}$ are partitioned such that

$$P_i^{(N-n)} = \left(x_{ir}^{(N-n)}, \qquad z_{ir}^{(N-n)}, \qquad x_{io}^{(n)}, z_{io}^{(n)} \right)$$

is an element of a set of partitions Γ relevant to a schema S of observational omissions where $(x_{ir}^{(N-n)}, z_{ir}^{(N-n)}, x_{io}^{(n)}, z_{io}^{(n)})$ represent $N - n$ retained and n omitted data sets so that $\Gamma = \Gamma(N, n, S)$. Let P = number of partitions of Γ.

3. Apply the predictive function to the retained observation set for each partition P_i and predict $x_{io}^{(n)}$ given $z_{io}^{(n)}$ by

$$\bar{x}_{io}^{(n)}(\omega) = \bar{x}_{io}\left(x_{ir}^{(N-n)}, \qquad z_{ir}^{(N-n)}, \qquad z_{io}^{(n)}, \omega \right).$$

4. Define a discrepancy measure or criterion function,

$$D_{N,n}(\omega) = (Pn)^{-1} \sum_{P_i \in \Gamma} d\left(x_{io}^{(n)}, \bar{x}_{io}^{(n)}(\omega) \right)$$

where $d(a, b)$ is a measure of the discrepancy between a and b.

5. Optimize (usually minimize) $D_{N,n}(\omega)$ w.r.t. ω and obtain $\hat{\omega}$. Form the predictor from all $x^{(N)}$ as

$$\hat{x}^{(n)} = x\left(x^{(N)}, z^{(N)}, z^{(n)}, \hat{\omega} \right).$$

If a relative scaling of predictive values is desirable one can compute the set of values $x(x^{(N)}, z^{(N)}, z^{(n)}, \omega)$ for all ω such that $1 \le D_{N,n}(\omega)/D_{N,n}(\hat{\omega}) \le A$.

Example 2.13. Let $h(x^{(N)})$ be a data based predictor of the next observation, g be some prior guess, and the predictive function be

$$x = \omega h(x^{(N)}) + (1 - \omega)g \qquad 0 \le \omega \le 1.$$

For a schema of one-at-a-time omissions let

$$D_{N,1}(\omega) = N^{-1} \sum_{j-1}^{N} \left(\omega h_j + (1 - \omega)g - x_j \right)^2$$

where

$$h_j = h\left(x_j^{(N-1)} \right)$$

is the same function h but based on all but x_j. Now

$$\frac{dD}{d\omega} = 0 = \sum_j \left(\omega h_j + (1 - \omega)g - x_j \right)\left(h_j - g \right)$$

$$= \omega \Sigma (h_j - g)^2 - \Sigma (x_j - g)(h_j - g)$$

so that an unrestricted solution is

$$\hat{\omega} = \frac{\Sigma(h_j - g)(x_j - g)}{\Sigma(h_j - g)^2}.$$

Hence employing the restriction on ω we obtain

$$\hat{x} = h(x^{(N)}), \qquad\qquad \text{if } \hat{\omega} \ge 1$$

$$= g \qquad\qquad\qquad\quad \text{if } \hat{\omega} \le 0$$

$$= \hat{\omega} h(x^{(N)}) + (1 - \hat{\omega})g \qquad \text{otherwise.}$$

If $h(x^{(N)}) = \bar{x}$ and $h_j = \bar{x}_{(j)}$, i.e., $\bar{x}_{(j)} = (N\bar{x} - x_j)/(N - 1)$ then, for $s^2 = (N - 1)^{-1} \Sigma (x_j - \bar{x})^2$

$$\hat{\omega} = \frac{t^2 - 1}{t^2 + (N-1)^{-1}} \qquad \text{where } t^2 = \frac{N(\bar{x} - g)^2}{s^2} > 1$$

$$\hat{\omega} = 0 \qquad\qquad \text{otherwise}$$

or

$$\hat{x} = \frac{(t^2 - 1)\bar{x} + N(N-1)^{-1}g}{t^2 + (N-1)^{-1}} \qquad t^2 > 1$$

$$\hat{x} = g \qquad\qquad t^2 \leq 1.$$

Note as t^2 increases, i.e., as $\bar{x}$ deviates further from the guess g relative to s^2, $\hat{x} \to \bar{x}$.

Example 2.14: *Application to analysis of variance models.* We consider the situation where we have J groups or treatments and K observations per group. Let x_{kj} be the kth observation of the jth group $k = 1, \ldots, K$ and $j = 1, \ldots, J$. The usual set of estimators of the group means, say $\theta_1, \ldots, \theta_J$ is

$$\bar{x}_{\cdot j} = K^{-1} \sum_k x_{kj} \qquad j = 1, \ldots, J$$

when the customary normality and homoscedasticity assumptions are made. It was noted by Stein (1962) from an inadmissibility point of view (and from a Bayesian point of view) that $(1 - \omega)\bar{x}_{\cdot j} + \omega \bar{x}_{\cdot \cdot}$ for $\bar{x}_{\cdot \cdot} = J^{-1} \Sigma_j \bar{x}_{\cdot j}$ and $0 \leq \omega \leq 1$, could yield a better set of estimates for $J \geq 3$ w.r.t. squared error. One such "Stein" estimator, which used

$$\omega_s = \min(1, C_1 m_1 / C_2 m_2)$$

where

$$C_1 = J(K-1)/\{J(K-1) + 2\}, \qquad C_2 = (J-1)/(J-3)$$

$$m_1 = J^{-1}(K-1)^{-1} \sum_k \sum_j (x_{kj} - \bar{x}_{\cdot j})^2$$

$$m_2 = K(J-1)^{-1} \sum_j (\bar{x}_{\cdot j} - \bar{x}_{\cdot \cdot})^2$$

rendered the usual set of estimators, i.e., $\omega = 0$ inadmissible. We shall use this convex combination to obtain predictors of the next observation in each group by two methods.

The first predictive method for evaluating alternative procedures, giving rise to predictors of the form $(1 - \omega)\bar{x}_{\cdot j} + \omega\bar{x}_{\cdot\cdot}$, involves omitting, say, the kth observation in the jth group, and computing the predictor from the $N - 1$ remaining observations. Here $N = KJ$, $K \geq 2$ and the predictor is

$$(1 - \omega)c_{kj} + \omega\bar{c}_{kj}$$

where

$$\bar{x}_{(kj)j} = c_{kj} = \left(K\bar{x}_{\cdot j} - x_{kj} \right)/(K - 1),$$

$$\bar{x}_{(kj)} = \bar{c}_{kj} = \left(N\bar{x}_{\cdot\cdot} - x_{kj} \right)/(N - 1).$$

This produces a predictor for x_{kj}, namely $(1 - \omega)c_{kj} + \omega\bar{c}_{kj}$. Then, repeating this for all k and j, we can compute the mean squared predictive discrepancy

$$s_{\omega}^2 = N^{-1} \sum_{j=1}^{J} \sum_{k=1}^{K} \left\{ (1 - \omega)c_{kj} + \omega\bar{c}_{kj} - x_{kj} \right\}^2.$$

The usual identities of the analysis of variance lead to

$$s_{\omega}^2 = \frac{K\{JK - 1 - \omega(J - 1)\}^2}{(K - 1)(JK - 1)^2} m_1 + \frac{\omega^2 JK(J - 1)}{(JK - 1)^2} m_2,$$

where m_1 and m_2 are as defined above. To obtain the optimal ω we minimize s_{ω}^2 with respect to ω. This yields

$$\hat{\omega} = \min\left(\frac{(JK - 1)m_1}{(J - 1)m_1 + (K - 1)Jm_2}, 1 \right).$$

As $J \to \infty$, $\omega \to (Km_1)/\{m_1 + (K - 1)m_2\}$, for fixed K. For fixed J, as $K \to \infty$, the estimator tends to $\min(m_1 m_2^{-1}, 1)$. In particular for $\omega = 0$

$$s_0^2 = \frac{K}{K-1} m_1.$$

Hence $(1 - \hat{\omega})\bar{x}_{.j} + \hat{\omega}\bar{x}_{..}$ is the set of predictors for a future set of observations from the J treatments. We shall refer to this as $\hat{\omega}_1$ to indicate it was derived via Method I in what follows.

Example 2.15: *The mixed model and method II.* We consider initially the mixed model and from it develop the method for the random effects case. Suppose that we have K vector observations each of J components. Assume that we are considering predictors of the components of a future $x_k' = (x_{k1}, \ldots, x_{kJ})$ of the form $(1 - \omega)\bar{x}_{.j} + \omega\bar{x}_{..}(j = 1, \ldots, J)$. Here in the mixed model there is a natural order, i.e., the kth vector $x_k' = (x_{k1}, \ldots, x_{kJ})$ is the fundamental sampling unit. Now we omit the kth row using $(1 - \omega)y_{kj} + \omega y_{kj}$, where

$$y_{kj} = \left(K\bar{x}_{.j} - x_{kj} \right) / (K - 1),$$

$$\bar{y}_{kj} = \left(KJ\bar{x}_{..} - \sum_{j=1}^{J} x_{kj} \right) \Big/ \{J(K-1)\},$$

to predict the missing row x_k'. Repeating this for $k = 1, \ldots, K$, we compute the mean squared predictive discrepancy

$$v_\omega^2 = J^{-1}K^{-1} \sum_{j=1}^{J} \sum_{k=1}^{K} \left\{ (1 - \omega) y_{kj} + \omega \bar{y}_{kj} - x_{kj} \right\}^2.$$

Again by simple algebra and the usual identities of the analysis of variance,

$$v_\omega^2 = \frac{(K-\omega)^2}{K(K-1)} m_1 + (JK)^{-1}(J-1)\omega^2 m_2 + \frac{\omega(2K-\omega)}{JK(K-1)} m_3,$$

where

$$m_3 = J(K-1)^{-1} \sum_{k=1}^{K} (\bar{x}_k. - \bar{x}..)^2, \qquad \bar{x}_k = J^{-1} \sum_{j=1}^{J} x_{kj}.$$

Further for $\omega = 0$, $v_0^2 = K(K-1)^{-1} m_1$ and minimization of v_ω^2 with respect to $\omega \in [0,1]$ yields $\min(\omega^*, 1)$, where

$$\omega^* = \min\left(\frac{K(m_1 J - m_3)}{(J-1)(K-1)m_2 + Jm_1 - m_3}, 1 \right) \geq 0$$

since $m_1 J \geq m_3$. Hence we have developed a method for predicting for the mixed model.

For the original situation, we note that the rows are not naturally associated. However, for this situation, the particular analysis for the mixed model is arbitrary, in terms of the row association. Now we need to consider the simultaneous omission of J observations one from each column in a nonarbitrary manner. This is accomplished by considering all possible permutations within each column and averaging v_ω^2 over all such possibilities. It is clear that m_1 and m_2 are unaffected by such permutations and that only m_3 need be averaged. This average can easily be shown to be m_1. Hence substitution in (3.3) yields

$$t_\omega^2 = \left\{ \frac{(K-\omega)^2 + J^{-1}(2K-\omega)\omega}{K(K-1)} \right\} m_1 + (JK)^{-1}(J-1)\omega^2 m_2.$$

Therefore the relative evaluation of particular ωs given a set of data is also possible by computing t_ω^2. For $\omega = 0$, $t_0^2 = K(K-1)^{-1} m_1$.

In this case minimization of t_ω^2 with respect to ω yields

$$\hat{\omega}_2 = \min\left(\frac{Km_1}{(K-1)m_2 + m_1}, 1 \right).$$

As K increases, the estimator tends to $\min(m_1 m_2^{-1}, 1)$. Note also

that

$$\hat{\omega}_2 \geq \hat{\omega}_1 \geq \min(m_1 m_2^{-1}, 1)$$

which implies greater shrinkage when a set of J observations is omitted as opposed to a single observation. The predictors from the full set of data for the prediction of a new set of observations one from each of the J groups is then $(1 - \hat{\omega}_2)\bar{x}_{.j} + \hat{\omega}_2 \bar{x}_{..}$ for $j = 1, \ldots, J$.

Note also that the predicted difference between a future observation from group j and one from group j' is $(1 - \hat{\omega}_i)(\bar{x}_{.j} - \bar{x}_{.j'})$.

Numerical illustration of Examples 2.14 and 2.15. To illustrate the shrinkage prediction methods, samples of size 10 were drawn from normal populations with means $(0.35, 0.1, -0.1, -0.35)$ and common variance 1 (Table 2.2).

Table 2.2   *Random samples of size 10 from* Π_i, $i = 1, \ldots, 4$

Π_1	0.118,	1.914,	1.659,	-0.175,	-0.216,
Π_2	1.99,	-0.375,	0.820,	0.528,	0.755,
Π_3	-0.304,	-0.018,	1.414,	-0.454,	0.015,
Π_4	1.141,	-0.140,	-1.412,	-0.709,	-0.968,
Π_1	0.105,	1.087,	2.331,	0.853,	-0.276
Π_2	-0.711,	-0.033,	1.484,	-0.593,	0.866
Π_3	2.081,	0.850,	-1.29,	0.260,	2.437
Π_4	-0.068,	0.082,	-0.950,	-0.223	0.647

The corresponding sample means were $(0.670, 0.475, 0.499, -0.260)$. Note that the sample means of Π_2 and Π_3 are in reverse order of the population means. The shrinkage proportion's ω resulting from the three proposed methods were all in the interval $(0.50, 0.55)$. The resulting predictions are plotted in Figure 2.3. Note that $\omega = 0$ uses the sample means as the predicted values and $\omega = 1$ yields the grand mean as predicted values for all four populations. Plots of the various discrepancy measures vs. ω are given in Figure 2.4.

Example 2.16: Flattening simple regression. Let the set of observations be

$$(x_j, z_j), \quad j = 1, \ldots, N.$$

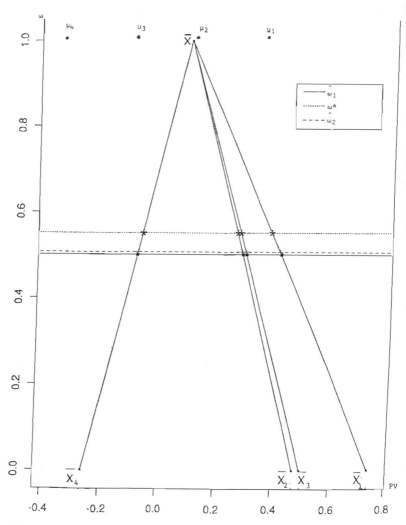

Figure 2.3   *Plot of predicted value (PV) vs. ω with true means μ $_i$ and sample means $\bar{x}_i$, i = 1,...,4 for various levels of shrinkage.*

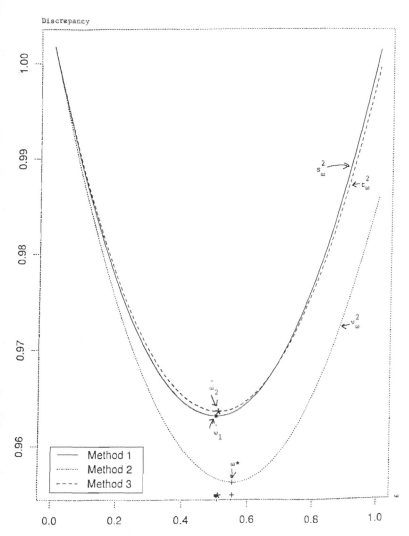

Figure 2.4   *Plots of discrepancy measures s_ω^2, v_ω^2, and t_ω^2 vs. ω.*

Specify the predictive function to be the convex combination of the mean and the least squares regression, namely,

$$x = (1 - \omega)\bar{x} + \omega[\bar{x} + b(z - \bar{z})] \qquad \omega \in [0, 1]$$
$$= \bar{x} + \omega b(z - \bar{z})$$

where

$$N\bar{z} = \Sigma z_j, \qquad N\bar{x} = \Sigma x_j, \qquad b = \Sigma(z_j - \bar{z})(x_j - \bar{x})/\Sigma(z_j - \bar{z})^2.$$

Assume squared discrepancy and one-at-a-time deletions so that

$$D(\omega) = N^{-1} \sum_j \left[\bar{x}_j + \omega b_j(z_j - \bar{z}_j) - x_j\right]^2$$

$$(N - 1)\bar{x}_j = N\bar{x} - x_j, \qquad (N - 1)\bar{z}_j = N\bar{z} - z_j$$

$$b_j = \sum_{i \neq j}(z_i - \bar{z}_j)(x_i - \bar{x}_j) \bigg/ \sum_{i \neq j}(z_i - \bar{z}_j)^2.$$

Minimization of $D(\omega)$ w.r.t. ω for $\omega \in [0, 1]$

$$\hat{\omega} = \begin{cases} 0 & \text{if } a \leq 0 \\ a \\ 1 & a \geq 1 \end{cases}$$

where

$$a = \sum_j (x_j - \bar{x})(z_j - \bar{z})b_j \bigg/ \sum_j (z_j - \bar{z})^2 b_j^2$$

yielding

$$\hat{x} = \bar{x} + \hat{\omega}b(z - \bar{z}).$$

For references see Geisser (1974, 1975) and Stone (1974).

2.4.5 Predictive sample reuse regions

We now present the setup for a predictive interval or region as described in Butler and Rothman (1980). It has the following ingredients.

i. A predictive region function:

$$\text{P.R.}(x^{(N)}; \omega) \qquad \text{for } x_{N+1}$$

ii. A criterion function: For simplicity assume one-at-a-time omissions.

$$D(\omega) = \frac{1}{N} \sum_{j=1}^{N} V\left\{\text{P.R.}\left(x_j^{(N-1)}; \omega\right)\right\}$$

where $V(\cdot)$ is defined as the volume of the jth region.

iii. A coverage frequency: We use a coverage frequency of at least $1 - \beta$ in a predictive simulation that is obtained for the given P.R. function. This is obtained by minimizing $D(\omega)$ w.r.t. ω subject to

$$\frac{1}{N} \sum_{j=1}^{N} I\left[x_j \notin \text{P.R.}\left\{x_j^{(N-1)}; \omega\right\}\right] \le \beta$$

where I is the indicator of the event in brackets, i.e., 1 if the event occurs and 0 otherwise.

iv. Predictive region: The solution is then used in

$$\text{P.R.}(x^{(N)}; \hat{\omega}).$$

Example 2.17. Let $[a]$ be the largest integer in a and

$$\text{P.R.}(x^{(N)}; \omega) = (y_\omega, y_{N-\omega+1}), \qquad \omega = 0, \dots, \left[\frac{N}{2}\right]$$

be the interval which depends on the symmetric order statistics, where y_k is the kth order statistic of the set $x_1, \dots, x_N$. Now omitting the jth observation

$$\text{P.R.}\left(x_j^{(N-1)}; \omega\right) = (y_{\omega.(j)}, y_{N-\omega+1.(j)})$$

$$= \begin{cases} [y_{\omega+1}, y_{N-\omega+1}] & \text{for } j = 1, \dots, \omega \\ [y_\omega, y_{N-\omega+1}] & \text{for } j = \omega+1, \dots, N-\omega \\ [y_\omega, y_{N-\omega}] & \text{for } j = N-\omega+1, \dots, N. \end{cases}$$

In this case the "volume" is the one-dimensional length $\mathscr{L}$. Hence

$$\mathscr{L}\{\text{P.R.}(x^{(N-1)};\omega\} = \begin{cases} y_{N-\omega+1} - y_{\omega+1} & \text{for } j = 1,\ldots,\omega \\ y_{N-\omega+1} - y_{\omega} & j = \omega+1,\ldots,N-\omega \\ y_{N-\omega} - y_{\omega} & j = N-\omega+1,\ldots,N \end{cases}$$

and

$$D(\omega) = \frac{1}{N}\left[\sum_{j=1}^{\omega}(y_{N-\omega+1}-y_{\omega+1}) + \sum_{j=\omega+1}^{N-\omega}(y_{N-\omega+1}-y_{\omega})\right.$$
$$\left. + \sum_{j=N-\omega+1}^{N}(y_{N-\omega}-y_{\omega})\right]$$
$$= \frac{1}{N}\left[\omega(y_{N-\omega+1}-y_{\omega+1}) + (N-2\omega)(y_{N-\omega+1}-y_{\omega})\right.$$
$$\left. + \omega(y_{N-\omega}-y_{\omega})\right]$$

Further $\text{Min}_{\omega} D(\omega)$ subject to

$$\frac{1}{N}\sum_{j=1}^{N} I\left[x_j \notin \text{P.R.}\left(x_j^{(n-1)};\omega\right)\right] \leq \beta$$

yields $\hat{\omega} = [N\beta/2]$ since $2\omega \leq N\beta$. Hence for $[a]$, the largest integer in a,

$$\text{P.R.}(x^{(N)};\hat{\omega}) = \left[y_{\left[\frac{N\beta}{2}\right]}, y_{N+1-\left[\frac{N\beta}{2}\right]}\right].$$

Note, for the case where $\beta = 2/N$ so that $\hat{\omega} = 1$ and the P.R. is (y_1, y_N), that the simulated relative coverage frequency is $(N-2)/N$ that x_{N+1} lies in (y_1, y_N). This is slightly less than $(N-1)/(N+1)$ obtainable on the assumption that the $x_1,\ldots,x_{N+1}$ are absolutely continuous and exchangeable (i.e., every permutation is equally likely and there are no ties). Here ties are permitted when they do not impede the calculation. Note that it is as if one lost a single observation because of this loosening of the stochastic structure.

References

Aitchison, J. (1975). Goodness of fit prediction. *Biometrika* **62**(3), 547–554.

Butler, R. (1986). Predictive likelihood inference with applications. *Journal of the Royal Statistical Society B* **48**, 1–38.

Butler, R., and Rothman, E. D. (1980). Predictive intervals based on reuse of the sample. *Journal of the American Statistical Association* **75**(372), 881–889.

Fisher, R. A. (1956). *Statistical Methods and Scientific Inference*. Edinburgh: Oliver and Boyd.

Geisser, S. (1956). A note on the normal distribution. *Annals of Mathematical Statistics* **27**, 858–859.

Geisser, S. (1974). A predictive approach to the random effect model. *Biometrika* **61**, 101–107.

Geisser, S. (1975). The predictive sample reuse method with applications. *Journal of the American Statistical Association* **70**, 320–328.

Hinkley, D. V. (1979). Predictive likelihood. *Annals of Statistics* **7**(4), 718–728.

Kalbfleisch, J. D. (1971). Likelihood methods of prediction. In *Foundations of Statistical Inference*, V. P. Godambe and D. A. Sprott (eds.). New York: Holt, Rinehart & Winston, 378–392.

Kullback, S., and Leibler, R. A. (1951). On information and sufficiency. *Annals of Mathematical Statistics* **22**, 79–86.

Lauritzen, S. L. (1974). Sufficiency, prediction and extreme models. *Scandinavian Journal of Statistics* **1**, 128–134.

Lejeune, M., and Faulkenberry, G. D. (1982). A simple predicted density function. *Journal of the American Statistical Association* **77**(379), 654–659.

Leonard, T. (1982). Comment. *Journal of the American Statistical Association*, **77**(379), 657–658.

Murray, G. D. (1977). A note on the estimation of probability density functions. *Biometrika* **64**, 150–152.

Rao, C. R. (1965). *Linear Statistical Inference and Its Applications*. New York: John Wiley.

Stein, C. (1962). Confidence sets for the mean of a multivariate normal distribution. *Journal of the Royal Statistical Society* **24**(2), 265–296.

Stone, M. (1974). Cross-validatory choice and assessment of statistical predictions. *Journal of the Royal Statistical Society B* **36**, 111–147.

CHAPTER 3

Bayesian prediction

As mentioned in Chapter 1, the ultimate Bayesian approach would be to completely model a set of observables and then calculate the probability distribution of the as yet unobserved values conditional on those already observed. With only a few exceptions this seems to be too burdensome a task. Hence for the most part we shall use the conventional Bayesian paradigm that models distributions of observables given parameters (or index set). One then can assign either subjective prior densities whenever possible (or appropriate) or useful so-called noninformative priors densities for the parameters of the statistical model. Once this is done, inferences or decisions about unobserved values conditional on those observed will flow from calculable probability distributions.

Before developing Bayesian prediction we offer three different interpretations of Bayes' (1763) work. In our view all of these interpretations to a greater or lesser degree can be viewed as predictivistic in character.

3.1 The "received version"

Let $X_1, \ldots, X_N$ be independent copies on

$$\Pr[X_i = 1 | \theta] = \theta = 1 - \Pr[X_i = 0 | \theta].$$

Let $R = \sum_{i=1}^{N} X_i$, then

$$\Pr[R = r | \theta] = \binom{N}{r} \theta^r (1 - \theta)^{N-r}.$$

This version, which has been traditionally accepted, avers that Bayes assumed that θ was a priori uniform in $(0, 1)$. Hence the posterior density of θ is

$$p(\theta|r) \propto \theta^r (1 - \theta)^{N-r}.$$

Price who communicated Bayes' posthumous essay calculated the predictive probability of the next binary variate as

$$\Pr[X_{N+1} = 1] = \frac{r + 1}{N + 2}.$$

It is possible that Bayes himself did this or discussed it with Price but we do not know for certain. This is a standard approach to Bayesian prediction. One thinks about and assumes a prior for the parameter θ and then calculates the posterior for θ. From this subsequently follows a predictive distribution for the unobserved values.

3.2   Revised version

Stigler (1982) claims that Bayes assumed that a priori

$$\Pr[R = r] = \frac{1}{N + 1} \qquad r = 0, 1, \ldots, N$$

and as before

$$\Pr[R = r|\theta] = \binom{N}{r} \theta^r (1 - \theta)^{N-r}.$$

Therefore for unknown $p(\theta)$

$$(N + 1)^{-1} = \int_0^1 p(\theta) \binom{N}{r} \theta^r (1 - \theta)^{N-r} \, d\theta$$

and $p(\theta) = 1$ is a solution to the above. Its uniqueness is assured if the moment generating function of $p(\theta)$ exists. Hence Bayes obtained without adequate proof

$$\bar{p}(\theta|r) \propto \theta^r(1-\theta)^{N-r}.$$

Here the prior is assumed on R the observed number of successes rather on the parameter θ. This approach, which assumes the conditional parametric model and formulates a prior based on an observable to attempt to find a prior on θ, appears more sensible in most situations but it cannot always be implemented, Geisser (1980b).

3.3   Stringent version—"compleat predictivist"

Another version (Geisser, 1985a) is now presented. In his essay Bayes imagined a ball being rolled on a unit square flat billiard table with the horizontal coordinate of the final resting place uniformly distributed in the unit interval. Call this random variable Y_0 and by assumption $f_{Y_0}(y) = 1$, $0 \le y \le 1$. A second ball is then rolled N times and we are informed of the number of times the second ball did not exceed the resting horizontal coordinate of the first ball $Y_0 = y$. The object is to use this information to infer (retrodict) the value y. Note that we have $Y_0, Y_1, \ldots, Y_N$ all i.i.d. and uniform and

$$X_i = \begin{cases} 1 & \text{if } Y_i \le y \\ 0 & \text{otherwise} \end{cases}.$$

Let $\Sigma X_i = R$ so that

$$\Pr[R = r|y] = \binom{N}{r} y^r(1-y)^{N-r}$$

and by Bayes' theorem

$$p(y|r) \propto y^r(1-y)^{N-r}$$

without the intrusion of parameters. This approach assumes only observables and models them. Then one calculates the conditional distribution of the unobserved given those observed.

3.4 Bayesian prediction format

As the parametric format is the simplest to work with we shall use it. Hence let

$$f\left(x^{(N)}, x_{(M)}|\theta\right) = f\left(x_{(M)}|x^{(N)}, \theta\right)f\left(x^{(N)}|\theta\right).$$

Let $\mathscr{L}(a, x_{(M)})$ denote the loss incurred in taking action $a(x^{(N)})$ $\in A$ on observing $X^{(N)} = x^{(N)}$ if future $X_{(M)} = x_{(M)}$ where A is the space of contemplated actions. Calculate

$$f\left(x_{(M)}, x^{(N)}\right) = \int f\left(x^{(N)}, x_{(M)}|\theta\right)p(\theta)\,d\theta$$

where $p(\theta)$ is the prior density and

$$f\left(x_{(M)}|x^{(N)}\right) = \frac{f\left(x_{(M)}, x^{(N)}\right)}{f\left(x^{(N)}\right)}$$

$$= \int f\left(x_{(M)}|\theta\right)p(\theta|x^{(N)})\,d\theta$$

where

$$p(\theta|x^{(N)}) \alpha f(x^{(N)}|\theta)p(\theta).$$

Further let

$$\mathscr{L}(a) = E\left[\mathscr{L}(a, x_{(M)})\right] = \int \mathscr{L}(a, x_{(M)})\,dF\left(x_{(M)}|x^{(N)}\right)$$

and choose that a^* such that it minimizes the loss or

$$\mathscr{L}(a^*) = \inf_a \mathscr{L}(a).$$

If the loss is additive and equal for each component of $x_{(M)}$ then

$$\mathscr{L}_M(a, x_{(M)}) = \sum_{i=1}^{M} \mathscr{L}(a, x_{N+i})$$

$$\mathscr{L}_M(a) = E\left[\sum_{i=1}^{M} \mathscr{L}(a, x_{N+i})\right]$$

$$= \sum_i \int \mathscr{L}(a, x_{N+i}) \, dF(x_{N+i}|x^{(N)})$$

$$= \sum_i E\mathscr{L}(a, x_{N+i}).$$

If $F(x_{(M)}|x^{(N)})$ is such that the marginal distribution of each component is identical, then

$$\mathscr{L}_M(a) = M\int \mathscr{L}(a, x) \, dF(x|x^{(N)}) = M\mathscr{L}_1(a)$$

so that the loss depends only on the common marginal predictive density.

   Example 3.1. Let A be a measurable set, then for $K > 0$ define the loss as

$$\mathscr{L}(A, x_{(M)}) = \begin{cases} \beta V_M(A) - K & \text{if } x_{(M)} \in A \\ \beta V_M(A) & \text{if } x_{(M)} \in A^C \end{cases}$$

where V_M is the Lebesgue measure in M dimensions (or the hypervolume). Hence for $f = f(x_{(M)}|x^{(N)})$

$$\mathscr{L}_M(A) = E\mathscr{L}_M(A, x_{(M)})$$

$$= \int_A [\beta V_M(A) - K] f \, dx_{(M)} + \int_{A^C} \beta V_M(A) f \, dx_{(M)}$$

$$= \int_{A\cup A^C} \beta V_M(A) f \, dx_{(M)} - K\int_A f \, dx_{(M)}$$

$$= \beta V_M(A) - K\int_A f \, dx_{(M)} = \beta \int_A dx_{(M)} - K\int_A f \, dx_{(M)}$$

$$\mathscr{L}_M(A) = \int_A (\beta - Kf) \, dx_{(M)}.$$

Hence

$$\text{Inf}_A \mathscr{L}_M(A) = \text{Inf}_A \int_A (\beta - Kf)\, dx_{(M)} = \mathscr{L}_M(A^*),$$

where clearly

$$A^* = \left\{ x_{(M)} : f \ge \frac{\beta}{K} \right\}.$$

If

$$\mathscr{L}_M(A, x_{(M)}) = \sum_{i=1}^{M} \mathscr{L}(B_i, x_{N+i})$$

for $A = B_1 \times B_2 \times \cdots \times B_M$, let

$$\mathscr{L}(B_i, x_{N+i}) = \begin{cases} \beta V(B_i) - K & \text{if } X_{N+i} \in B_i \\ \beta V(B_i) & \text{if } X_{N+i} \in B_i^C. \end{cases}$$

Further, if $f(x_{(M)}|x^{(N)})$ is such that all its components have the same marginal density, and $B_i \equiv B$ so that $A = B \times B \cdots \times B$, then

$$\mathscr{L}_M(A) = M\mathscr{L}(B).$$

Hence

$$\inf_A \mathscr{L}_M(A) = \mathscr{L}(A^*) = M\mathscr{L}(B^*)$$

where $A^* = B^* \times B^* \times \cdots \times B^*$ and

$$B^* = \left\{ x : f(x|x^{(N)}) \ge \frac{\beta}{K} \right\}.$$

This example provides a basis for the use of highest probability density regions for prediction.

3.5   Prior distributions

In situations where θ is a meaningful physical entity (say measuring a physical object) then one may think about a subjective distribution for the entity. In many other cases one is using $f(x|\theta)$ as a convenient modeling device. Here the existence of θ is problematical except perhaps as a limiting function of observables [this will happen when the assumption of infinite exchangeability is made, i.e., the joint density of $(X_1, \ldots, X_n)$ for all n is completely permutable] and, more importantly, inferential focus is constrained to observables and not to θ. From this point of view subjectivity should be directed at the distribution of observables themselves. This would entail assigning a subjective distribution directly to $F(x^{(N)}, x_{(M)})$ from which we would calculate $F(x_{(M)}|x^{(N)})$. However, this appears to be too burdensome to be worthwhile except in certain limited situations as in the following example.

Example 3.2. Suppose there is a sequence of binary events, say success and failure, $X_1, \ldots, X_N, X_{N+1}, \ldots, X_{N+M}$. We shall observe N of them and wish to predict the total number of future successes of the M unobserved events. Out of the N drawn we observe $T = t$ successes. Hence, for $R = r$ the future number of successes and $S = T + R$,

$$\Pr(T = t|s) = \binom{s}{t}\binom{N+M-s}{N-t}\bigg/\binom{N+M}{N},$$

$$t = 0, 1, \ldots, \min(N, s).$$

If we assume prior to any sampling that the total number of successes S is equally likely to be any number $S = 0, 1, \ldots, N + M$, then

$$\Pr[S = s|N + M] = (N + M + 1)^{-1}.$$

Since $S = T + R$, then

$$\Pr[R = r|N, M, t] = \binom{t+r}{t}\binom{N+M-t-r}{N-t}\bigg/\binom{N+M+1}{N+1},$$

$$r = 0, 1, \ldots, M.$$

3.6 Uniform priors

In the absence of real prior information of one kind or another and where a prior density for θ is required that reflects some initial ignorance of θ a method has been suggested by Jeffreys (1946). Initially he suggested that a parameter ranging over a finite interval or the whole real line should be uniformly distributed and if it ranged over a semiinfinite interval its logarithm should be uniform. For N independent copies where the likelihood is

$$L = \prod_1^N f(x_i | \theta),$$

the expected Fisher information quantity for scalar θ, when it exists, is

$$I_N(\theta) = E\left(\frac{d \log L}{d\theta}\right)^2.$$

Jeffreys later suggested that

$$p(\theta) \propto I^{1/2}(\theta)$$

where

$$I_N(\theta) = NI(\theta)$$

because of its invariance properties. For example

$$I(\theta) = I[\tau(\theta)]\left(\frac{d\tau(\theta)}{d\theta}\right)^2$$

or taking the positive square roots of both sides,

$$I^{1/2}(\theta) = I^{1/2}(\tau)\left|\frac{d\tau}{d\theta}\right|$$

or

$$I^{1/2}(\theta)\, d\theta = I^{1/2}(\tau)\, d\tau$$

where τ is differentiable and is continuous and nonzero for all but a finite number of values of θ.

Note that if we set

$$\tau = \int_{-\infty}^{\theta} I^{1/2}(\theta')\, d\theta'$$

then

$$d\tau = I^{1/2}(\theta)\, d\theta$$

and τ is uniformly distributed. Of course in a particular technical sense uniform distributions over an infinite range do not exist. However, using these improper priors will often lead to useful results. Another view is that these priors are locally uniform where the likelihood is large and essentially zero everywhere else. This point of view is set forth in Box and Tiao (1973) in terms of data translated likelihoods. They point out that if

$$f(x^{(N)}|\theta) = L\left[\tau - t(x^{(N)})\right]$$

then the shape and form of the likelihood for τ is completely known and identical except for a translation. Hence the data serve only to change the location of L depending on t and a locally uniform prior on τ will minimally affect the likelihood. This is useful when little or no information concerning τ exists a priori.

Hence

$$p(\tau)\, d\tau \propto d\tau.$$

Now this concords with Jeffreys' criterion since

$$I(\tau) = E\left(\frac{d \log L(\tau - t)}{d\tau}\right)^2 = E\left(\frac{d \log L(\tau - t)}{d(\tau - t)}\right)^2$$

so that

$$I^{1/2}(\tau) = \text{const.}$$

Then by our previous result regarding the transformation of $I(\theta)$

into $I(\tau)$,

$$p(\theta) = I^{1/2}(\theta) = \left| \frac{d\tau}{d\theta} \right|.$$

Example 3.3. Let

$$X_i \sim N(\mu, \sigma^2)$$

then

$$L(\mu) = \frac{1}{\sigma^N} e^{-(1/2\sigma^2)[(N-1)s^2 + N(\mu - \bar{x})^2]} \propto e^{-(N/2\sigma^2)(\mu - \bar{x})^2}$$

is data translated for known σ^2. Similarly, for known μ, $L(\log \sigma^2)$ is data translated and agrees with Jeffreys invariance criterion that μ is uniform given σ and $\log \sigma$ is uniform given μ.

However, not all $L(\theta)$ can be exactly data translated. If we assume some regularity conditions such that the maximum likelihood estimator $\hat{\theta}$ is subject to

$$(\hat{\theta} - \theta)\sqrt{N} \to N[0, I^{-1}(\theta)],$$

then using a Taylor series approximation,

$$\log L(\theta) \doteq \log L(\hat{\theta}) + \frac{(\theta - \hat{\theta})^2}{2} \left(\frac{d^2 \log L}{d\theta^2} \right)_{\theta = \hat{\theta}}$$

$$\log L(\theta) \doteq \log L(\hat{\theta}) - \frac{N}{2}(\theta - \hat{\theta})^2 \left(\frac{-1}{N} \frac{d^2 \log L}{d\theta^2} \right)_{\theta = \hat{\theta}}.$$

Suppose

$$J(\hat{\theta}) = -\frac{1}{N} \left(\frac{d^2 \log L}{d\theta^2} \right)_{\theta = \hat{\theta}}$$

depends only on $\hat{\theta}$, so that for varying θ

$$\log L \doteq C_1 - \frac{N}{2}(\theta - \hat{\theta})^2 J(\hat{\theta}),$$

and hence is approximately the log of a normal likelihood with variance $N^{-1}J^{-1}(\hat{\theta})$. If $J(\hat{\theta})$ is not constant the shape will vary. In this case let $\tau = \tau(\theta)$ be a one-to-one transform so that $\hat{\tau} = \tau(\hat{\theta})$ is as well. Then

$$J(\hat{\tau}) = -\frac{1}{N}\frac{d^2 \log L}{d\tau^2}\bigg|_{\tau=\hat{\tau}} = J(\hat{\theta})\left(\frac{d\theta}{\partial \tau}\right)^2_{\theta=\hat{\theta}}.$$

Hence if $\tau(\hat{\theta})$ is so chosen so that $J(\hat{\tau})$ is independent of $\hat{\tau}$, i.e.,

$$J^{1/2}(\hat{\tau}) = J^{1/2}(\hat{\theta})\left.\frac{d\theta}{d\tau}\right|_{\theta=\hat{\theta}} = C_2$$

or

$$J^{1/2}(\hat{\theta}) \propto \left.\frac{d\tau}{d\theta}\right|_{\theta=\hat{\theta}}$$

or

$$\tau = \int_{-\infty}^{\theta} J^{1/2}(\theta')\, d\theta'$$

$$\log L(\tau) \doteq C_1 - \frac{N}{2}(\tau - \hat{\tau})^2 C_2$$

i.e., approximately a normal likelihood with a constant variance so that the above is approximately data translated.

Note also if $J(\hat{\theta}, x^{(N)})$ is not a function of $\hat{\theta}$ alone, then as N grows

$$\frac{1}{N}\frac{d^2 \log L}{d\theta^2}\bigg|_{\theta=\hat{\theta}} = \frac{1}{N}\sum_{1}^{N}\frac{d^2 \log f(x|\theta)}{d\theta^2}\bigg|_{\theta=\hat{\theta}}$$

$$\rightarrow E\left[\frac{1}{N}\sum_{1}^{N}\frac{d^2 \log f(x_i|\theta)}{d\theta^2}\right] = -I(\theta)$$

by the strong law of large numbers so that

$$J(\hat{\theta}, x^{(N)}) \rightarrow I(\theta)$$

for large samples demonstrating the large sample data translating property of Jeffreys' rule.

For multiparameter situations, let

$$
\theta = \begin{pmatrix} \theta_1 \\ \vdots \\ \theta_p \end{pmatrix}, \qquad
\frac{\partial \log f}{\partial \theta} = \begin{pmatrix} \dfrac{\partial \log f}{\partial \theta_1} \\ \vdots \\ \dfrac{\partial \log f}{\partial \theta_p} \end{pmatrix}.
$$

The expected information matrix is

$$
\mathbf{I}(\theta) = E\left(\frac{\partial \log f}{\partial \theta} \right)\left(\frac{\partial \log f}{\partial \theta} \right)' = \frac{1}{N} E\left(\frac{\partial \log L(\theta)}{\partial \theta} \right)\left(\frac{\partial \log L(\theta)}{\partial \theta} \right)'.
$$

Here Jeffreys suggests using as prior density for θ

$$
p(\theta) \propto |\mathbf{I}(\theta)|^{1/2}.
$$

If

$$
f(x^{(N)}|\theta) = L\left[\tau - t(x^{(N)}) \right]
$$

where $L(\cdot)$ is a known function independent of $x^{(N)}$ and

$$
\tau(\theta) = \begin{pmatrix} \tau_1 \\ \vdots \\ \tau_p \end{pmatrix}
$$

is a one-to-one transformation θ and $t = (t_1, \ldots, t_p)$ are p functions of $x^{(N)}$, then $\mathbf{I}(\tau)$ is independent of τ. Hence using $p(\tau) \propto |\mathbf{I}(\tau)|^{1/2} = \text{const.}$ yields a locally uniform prior for τ. Now for the matrix of values

$$
\left(\frac{\partial \tau}{\partial \theta} \right) = \left(\frac{\partial \tau_i}{\partial \theta_j} \right) \qquad i, j = 1, \ldots, p
$$

we obtain

$$I(\theta) = \left(\frac{\partial \tau}{\partial \theta}\right)' I(\tau) \left(\frac{\partial \tau}{\partial \theta}\right).$$

Then

$$p(\theta) \propto |I(\theta)|^{1/2} = |I(\tau)|^{1/2} \left|\frac{\partial \tau}{\partial \theta}\right| \propto p(\tau) \left|\frac{\partial \tau}{\partial \theta}\right|$$

since $p(\tau) = p(\theta)|\partial\theta/\partial\tau|$, demonstrating the invariance.

Although exact data translated likelihoods do not frequently occur, we now assume regularity conditions for the maximum likelihood estimator $\hat{\theta}$ hold that lead to

$$\log L(\theta) \doteq \log L(\hat{\theta}) - \frac{N}{2}(\theta - \hat{\theta})' J(\hat{\theta})(\theta - \hat{\theta})$$

where the matrix

$$J(\hat{\theta}) = \left\{-\frac{1}{N}\frac{\partial^2 \log L}{\partial\theta_i\,\partial\theta_j}\right\}_{\theta - \hat{\theta}} \qquad i, j = 1,\ldots,p.$$

Again if $J(\hat{\theta})$ is independent of $\hat{\theta}$ we have an approximate data translated likelihood.

For large N even if $J(\hat{\theta}, x^{(N)})$ does not depend solely on $\hat{\theta}$ it is closely approximated by

$$J(\hat{\theta}) \doteq I(\hat{\theta}) = \frac{1}{N}E\left(\frac{\partial \log L}{\partial\theta}\right)\left(\frac{\partial \log L}{\partial\theta}\right)'\Big|_{\theta = \hat{\theta}} \to I(\theta).$$

We note that the approximation is actually an identity for the exponential class. This may vary with $\hat{\theta}$. One could seek $\tau = \tau(\theta)$ such that $I(\tau)$, independent of $\hat{\tau}$. This is not always possible. We can, however, relax the usual data translating requirement to make the hypervolume of the region of τ

$$(\tau - \hat{\tau})' I(\tau)(\hat{\tau} - \hat{\tau}) < \text{Const.}$$

invariant irrespective of $\hat{\tau}$. The volume of this hyperellipsoid is proportional to

$$|\mathbf{I}(\hat{\tau})|^{-1/2}.$$

So if a transformation can make $|\mathbf{I}(\hat{\tau})|$ independent of $\hat{\tau}$ we have a reasonable surrogate for a data translated likelihood. Hence

$$|\mathbf{I}(\tau)|^{1/2} = \text{Const.}$$

will be satisfied if

$$\left|\frac{\partial \tau}{\partial \theta}\right| \propto |\mathbf{I}(\theta)|^{1/2}$$

is the transformation. Data translated likelihoods and their variations were introduced by Box and Tiao (1973).

Several other approaches that produce reference priors that purportedly reflect knowing little a priori have been put forth by Akaike (1978), Bernardo (1979), and Zellner (1977).

There is yet another approach that stresses prediction and is in the spirit of a loss function approach that is used for estimating the sampling density in Chapter 2, Section 1.2.

We assume that in obtaining a predictive probability function for X_{N+1} that is close to the sampling density $f(x_{N+1}|\theta)$ we admit only certain forms of $f(x_{N+1}|x^{(N)})$. This in turn restricts the class C of admissible prior densities $p(\theta) \in C$. We then obtain that $p(\theta)$ that minimizes

$$E_{X^{(N)}} E_{X_{N+1}} \left\{ \log \frac{f(x_{N+1}|\theta)}{f(x_{N+1}|x^{(N)})} \right\}$$

with respect to the class C, if it exists, Geisser (1979).

3.7 Comments

3.7.1 *Arguments concerning the use of a noninformative prior*

There have been several major criticisms concerning the use of noninformative or reference priors. A point is made of the fact that the prior is often an improper density. This is not a particularly serious objection since one can consider the relatively noninformative prior as representing local behavior where the likelihood is appreciable and zero elsewhere (so that technically it is not improper but for ease of calculation it is used as an approximation). A more serious criticism is that a noninformative prior depends on the likelihood $L(\theta)$ so it is not really independent prior information, i.e., prior ignorance about something should not depend on the particular experiment. A reasonable response to this, enunciated by Box and Tiao (1973), is that although there is presumably no such thing as complete ignorance, the amount of information using these priors is small relative to what the experiment can provide. Usually an experiment is undertaken only when it can provide substantially more information than what is known beforehand (a priori), i.e., knowing little is specific to a particular experiment.

The most critical objection is that these noninformative priors do not generally represent true subjective knowledge. There are two points that mitigate against this objection in most cases. First, scientific investigations are considerably more convincing when strong subjective views of the experiment are withheld and data analyzed against some neutral or reference prior. Secondly, a model indexed by θ is obviously an approximation and often θ has no reality but exists only as an aid to modeling for predictive purposes. Hence a subjective prior about θ as an index may have no intrinsic relevance.

3.7.2 Why Bayes? The coherent posture

Clearly the major advantage of the Bayes approach for those who accept it is the capability to produce probability statements regarding the entities under scrutiny whether parameters or observables. However, it could be argued that these probability statements are devoid of objective reality. Be that as it may there is another argument that is compelling for many that induces the Bayesian approach and yields consistency in inference and decision making in a particular setting.

It is termed coherence and the metaphor it is most effectively expressed in is betting strategies (de Finetti, 1937). In its simplest form as given by Cornfield (1969) we suppose that we are dealing with a known discrete probability distribution

$$\Pr\left[X = x_i | \theta = \theta_j \right] = p_{ij} > 0 \qquad i = 1, \ldots, n$$

and a finite number of possibilities for the parameter $\theta = \theta_j$, $j = 1, \ldots, N$. A master of ceremonies chooses a particular θ_j and draws an observation on the random variable X given that θ_j and announces that he has obtained a particular x_i. Statistician B is then required to make probability assignments P_{ik} for all 2^N possible subsets I_k, $k = 1, \ldots, 2^N$, of $(\theta_1, \ldots, \theta_N)$. If θ_j, chosen by the master of ceremonies, is included in a subset it is deemed correct and incorrect if it isn't. An antagonist A may bet for or against any combination of the 2^N subsets given the same knowledge as B. Now B assigns P_{ik} and A assigns an amount S_{ik} that θ_j lies in I_k. A then gives an amount $P_{ik}S_{ik}$ to B and receives S_{ik} if θ_j lies in I_k, but receives nothing if it does not. Basically A is risking $P_{ik}S_{ik}$ to win $(1 - P_{ik})S_{ik}$. B's gain on this bet, when θ_j is chosen and x_i is realized, is

$$G_{ijk} = (P_{ik} - \delta_{jk})S_{ik}$$

where

$$\delta_{jk} = \begin{cases} 1 & \text{if } \theta_j \in I_k \\ 0 & \text{otherwise.} \end{cases}$$

Now summing over all possible subsets we get

$$G_{ij\cdot} = \sum_{k=1}^{2^N} G_{ijk} = \sum_k (P_{ik} - \delta_{jk})S_{ik}.$$

Further the expected gain for B when $\theta = \theta_j$ is

$$G_{\cdot j\cdot} = \sum_{i=1}^{n} p_{ij} \sum_{k=1}^{2^N} (P_{ik} - \delta_{jk})S_{ik}.$$

Now if $G_{\cdot j\cdot} \leq 0$ for any S_{ik} for all j and $G.j. < 0$ for at least one θ_j, we say that B's probability assignment is incoherent since on average he will lose money for at least one θ_j (or more) and at best hold his own for others. Otherwise we say he is coherent. Now if the P_{ik} are assigned in such a manner that

$$\Pr\left[\theta_j | x_i\right] = P_{ij} \propto q_j p_{ij} \qquad \text{for } q_j > 0$$

$$\sum_{j=1}^{N} P_{ij} = 1 \qquad \text{for } i = 1, 2, \ldots, n,$$

we say this is a Bayesian assignment (note the P_{ij}s are special cases of P_{ik} because all of the latter can be derived from the P_{ij}s). This leads to the following result.

Theorem. A Bayesian assignment of the probabilities P_{ik} by B is both necessary and sufficient for coherence.

A sketch of a proof using elementary methods is given by Cornfield (1969). Although in the previous discussion θ was termed a parameter (greek letter) it could also have been designated as an observable and given a roman letter (say, y). In this instance the odds and states would refer to a retrodiction—an event that has already occurred. There is also a proof of this result by Freedman and Purves (1969) that is extended for prediction using the separation theorem for convex sets.

In summary, this theorem implies that if B's betting odds on the subsets are not consistent with prior probabilities, then A, by properly choosing the stakes S_{ik}, can disadvantage B financially or in whatever utility units are expressed by the stakes S_{ik}.

Theorems regarding coherence in a more general setting formulated directly for predictive inference have been derived by Lane and Sudderth (1984). This paper also gives some guidance as to when the use of improper priors will lead to coherent predictions (proper priors will always do so).

While there is agreement as to the consistency of the Bayesian approach there remains considerable controversy over its implementation and usefulness.

3.8   Conjugate priors

Conjugate priors or priors closed under sampling are another convenient framework. Here the prior information is approximately presumed to be such that when combined with the likelihood the posterior and the prior belong to the same family of distributions. Let $p(\theta|\tau)$ be the prior probability function of θ given a set of hyperparameters τ. The requirement that $p(\theta|\tau)$ and $p(\theta|x^{(N)}, \tau)$ belong to the same family for all N implies that for given τ the marginal predictive probability function of $X_{(M)}$ and $X_{(M)}|x^{(N)}$ will also belong to the same family. The marginal of $x^{(N)}$ can be useful for estimating τ when it is unknown (Morris, 1983). If a reasonable estimate of τ is obtained, it can be inserted into the predictive distribution of $X_{(M)}$ given $x^{(N)}$ and τ to yield an approximate predictive distribution.

Distributions closed under sampling were first introduced by Barnard (1954) and later were called conjugate distributions by Raiffa and Schlaifer (1961) who promulgated their use. For sufficient conditions that guarantee a conjugate family see De Groot (1970).

Example 3.4. Let $X_1, \ldots, X_{N+M}$ be independent copies on

$$f(x|\theta) = \theta e^{-\theta x} \qquad \theta > 0, x \geq 0.$$

Assume

$$p(\theta) \propto \theta^{\delta-1} e^{-\gamma \theta} \qquad \gamma > 0, \delta > 0$$

and $\tau = (\gamma, \delta)$. Hence the predictive density of $X_{(M)}$ is easily calculated as follows:

$$p(\theta|x^{(N)}) \propto \theta^{N+\delta-1} e^{-\theta(\gamma + N\bar{x})}$$

$$f(x_{(M)}|x^{(N)}, \tau) = \int f(x_{(M)}|\theta) p(\theta|x^{(N)}) \, d\theta$$

$$= \frac{\Gamma(N+M+\delta)}{\Gamma(N+\delta)}$$

$$\times \frac{(\gamma + N\bar{x})^{N+\delta}}{(\gamma + N\bar{x} + x_{N+1} + \cdots + x_{N+M})^{N+M+\delta}}.$$

Note that the marginal density $f(x_{(M)}|\tau)$ has the same form as the above with $N = 0$. While the predictive density will exist for the noninformative case, i.e., $\delta = \gamma = 0$, the marginal will not. For

$M = 1$, we obtain

$$f(x_{N+1}|x^{(N)}, \tau) = \frac{(N + \delta)(N\bar{x} + \gamma)^{N+\delta}}{(N\bar{x} + \gamma + x_{N+1})^{N+\delta+1}}. \qquad (3.1)$$

Further

$$E(X_{N+1}|x^{(N)}, \tau) = \frac{\gamma + N\bar{x}}{N + \delta - 1}$$

$$\text{Mode } (X_{N+1}) = 0.$$

Since the predictive density is decreasing in $(0, \infty)$ then an HPD interval is defined by

$$\frac{(N + \delta)(N\bar{x} + \gamma)^{N+\delta}}{(N\bar{x} + \gamma + x_{N+1})^{N+\delta+1}} \geq \frac{\beta}{K}$$

or

$$0 \leq X_{N+1} \leq \left(\frac{N + \delta}{\beta/K}\right)^{1/(N+\delta+1)} (N\bar{x} + \gamma)^{(N+\delta)/(N+\delta+1)} - N\bar{x} - \gamma$$

$$= x_{1-\alpha}.$$

The probability content of the interval is

$$\Pr[X_{N+1} < x_{1-\alpha}|x^{(N)}, \tau] = 1 - \alpha = 1 - \left(1 + \frac{x_{1-\alpha}}{N\bar{x} + \gamma}\right)^{-(N+\delta)}.$$

If we fix α we can solve for β/K. Note that here we assumed that our subjective assessment for θ could be expressed as a gamma prior

$$p(\theta) \propto \theta^{\delta-1} e^{-\gamma\theta}$$

with γ and δ known. Suppose δ and γ are not assumed known, what alternatives are there? One can put distributions on these hyperparameters as well which may or may not depend on further hyperparameters. This can go on indefinitely and where does one stop? If γ and δ can be assigned completely known distributions we stop. If not, perhaps we can substitute estimates for γ and δ that will not alter our inferences in any appreciable manner.

3.9 Approximate Bayes procedures

While it is true that all procedures are Bayesian approximations of one sort or the other, essentially they vary as to the degree of uncertainty about particular components of the model. In particular, in hierarchical models we discuss methods of using estimates for unknown hyperparameters.

Some of the possible methods come under the heading of Empirical Bayes procedures but it is better put to call them approximations for the hyperparameters. Hence we would obtain an approximate predictive distribution

$$\hat{F}\left(x_{(M)}|x^{(N)}, \tau\right) = F\left(x_{(M)}|x^{(N)}, \hat{\tau}\right).$$

Methods for approximating the hyperparameters have been suggested that depend on calculating the marginal density (or likelihood)

$$f\left(x^{(N)}|\tau\right) = \int f\left(x^{(N)}|\theta\right) p(\theta|\tau)\, d\theta$$

and applying either of two usual estimation procedures to this marginal likelihood, namely:

1. Maximum likelihood estimation.
2. Method of Moments.

Another procedure that uses characteristics of the predictive distribution itself, namely the Predictive Sample Reuse method, can also be applied.

Example 3.4 (*continued*). For maximum likelihood estimation we first calculate

$$f\left(x^{(N)}|\tau\right) = \frac{\Gamma(N+\delta)\gamma^{\delta}}{\Gamma(\delta)(N\bar{x}+\gamma)^{N+\delta}}$$

or

$$\log f = \log \frac{\Gamma(N+\delta)}{\Gamma(\delta)} + \delta \log \gamma - (N+\delta)\log(N\bar{x}+\gamma)$$

and

$$\frac{d\log f}{d\delta} = \sum_{j=1}^{N} \frac{1}{N+\delta-j} + \log \gamma - \log(N\bar{x}+\gamma) = 0$$

$$\frac{d\log f}{d\gamma} = \frac{\delta}{\gamma} - \frac{N+\delta}{N\bar{x}+\gamma} = 0.$$

From the latter, $\gamma = \delta\bar{x}$. Now if δ were assumed known, the derivative with respect to δ would be unnecessary. Here the solution would substitute $\delta\bar{x}$ for γ in the predictive density. This effectively increases the sample size from N to $N+\delta$, an acceptable result. However, if we are required to estimate δ as well, then substituting $\gamma = \delta\bar{x}$ in the aforementioned derivative yields

$$\sum_{j=1}^{N} \frac{1}{N+\delta-j} = \log\frac{N+\delta}{\delta}.$$

Clearly the solution here is $\delta = \infty$ and then $\hat{\gamma} = \infty$ since $\bar{x} > 0$ with probability 1. It is then clear $\hat{\gamma} = \hat{\delta}\bar{x}$ and $\hat{\delta} = \infty$ maximizes $\log f$. Hence as $\delta = \infty$ and $N \geq 1$,

$$\Pr(\theta = \bar{x}^{-1}) = 1$$

and

$$f\left(x_{N+1}|x^{(N)}\right) = \frac{1}{\bar{x}} e^{-x_{N+1}/\bar{x}},$$

irrespective of N. This is unacceptable since this merely substitutes the maximum likelihood estimator of θ, $\hat{\theta} = \bar{x}^{-1}$ in the predictive density of x_{N+1} for all N. The estimate above would only be adequate for sufficiently large N. Clearly the diffuseness of the predictive density of X_{N+1} should depend on N, and diminish as N increases. The above belies this desideratum.

For application of the method of moments we first need to equate the first two sample moments of the exchangeable random variables $X_1, \ldots, X_N$ in $f(x^{(N)}|\delta, \gamma)$ to their expectations. This yields

$$\bar{x} = \frac{\gamma}{\delta - 1}$$

and

$$N^{-1} \sum_1^N x_i^2 = \frac{2\gamma^2}{(\delta - 1)(\delta - 2)}$$

with the restriction that $\delta > 2$. Note that originally δ was only restricted to be positive. The solution here is, for $N \geq 2$,

$$\delta_M = \max\left[2, \frac{2(N-1)}{N-1-t^2}\right]$$

where $t^2 = N\bar{x}^2/s^2$ and $(N-1)s^2 = \sum_1^N (x_i - \bar{x})^2$, and

$$\gamma_M = \bar{x}(\delta_M - 1) \geq \bar{x}.$$

A drawback is the restriction $\delta_M \geq 2$ which is minor when compared to $\hat{\delta} \to \infty$ for the maximum likelihood estimator. Hence the estimated predictive density is

$$f_M(x_{N+1}|x^{(N)}) = \frac{(N+\delta_M)(N\bar{x}+\gamma_M)^{N+\delta_M}}{(N\bar{x}+\gamma_M+x_{N+1})^{N+\delta_M+1}}.$$

We noted previously that for unbounded values of (γ_M, δ_M), which resulted from the maximum likelihood values for example, we obtained

$$\frac{1}{\bar{x}} e^{-x_{N+1}/\bar{x}}$$

and as N grows this approaches $\theta e^{-\theta x_{N+1}}$. Hence, whether or not we obtain finite or unbounded values for (δ_M, γ_M), as N grows,

$$f_M\left(x_{N+1}|x^{(N)}\right) \to \theta e^{-\theta x_{N+1}}.$$

However it can be shown that for all N, γ_M and δ_m are finite with probability 1. At any rate, it would appear that the method of moments is a definite improvement over the maximum likelihood approach for this problem.

We now estimate the predictive density based on the predictive sample reuse (PSR) approach. In this case we do not use the marginal density $f(x^{(N)}|\delta, \gamma)$ but the predictive expectation

$$E\left(X_{N+1}|x^{(N)}, \delta, \gamma\right) = \frac{\gamma + N\bar{x}}{N + \delta - 1},$$

from the actual predictive density of X_{N+1}. We can form a predictor for x_j from $x_{(j)}$, the other $N-1$ observations with x_j deleted, along with γ and δ based on the above expectation, namely

$$\bar{x}_j = \frac{\gamma + (N-1)\bar{x}_{(j)}}{N - 1 + \delta - 1}$$

where $(N-1)\bar{x}_{(j)} = \sum_{i \neq j}^{N} x_i$. We then form a discrepancy measure

$$D = \sum_j \left(x_j - \frac{\gamma + (N-1)\bar{x}_{(j)}}{N - 1 + \delta - 1}\right)^2$$

and minimize this with respect to γ and δ. Taking derivatives of D with respect to γ and δ and setting them equal to zero yields as solutions

$$\tilde{\delta} = \max\left[1, \frac{t^2 + (N-1)/(N-2)}{t^2 - (N-1)/(N-2)}\right]$$

$$\tilde{\gamma} = (\tilde{\delta} - 1)\bar{x}.$$

Note that as t^2 grows $\tilde{\delta} \to 1$ from above and $\tilde{\gamma} \to 0$ from above. Hence the estimate of δ is at least as large as 1, which is an improvement over the method of moments approach which required the estimate of δ to be at least as large as 2. The estimated predictive density now is

$$\bar{f}\left(x_{N+1}|x^{(N)}\right) = \frac{(N+\bar{\delta})(N\bar{x}+\bar{\gamma})^{N+\bar{\delta}}}{(N\bar{x}+\bar{\gamma}+x_{N-1})^{N+\bar{\delta}+1}}$$

and we note that as N grows $\bar{f}(x_{N+1}|x^{(N)})$ approaches the sampling density of X_{N+1} since $(\bar{\delta}, \bar{\gamma})$ will be finite with probability 1 for all N. Using the relationship

$$\gamma = \bar{x}(\delta - 1)$$

we express the method of moments and sample reuse estimated predictive density as

$$\frac{(N+\delta)}{(N+\delta-1)\bar{x}}\left(1 + \frac{x_{N+1}}{(N+\delta-1)\bar{x}}\right)^{-(N+\delta+1)} \tag{3.2}$$

with $\delta = \bar{\delta}$ or δ_M, and even for $\delta \to \infty$, the MLE estimator. Note however that for the non-informative improper prior density on θ, namely,

$$p(\theta) \propto \frac{1}{\theta}$$

we substitute $\delta = \gamma = 0$ in (3.1) not in (3.2) to attain the appropriate result. All of these values for δ in (3.2) serve only to change the effective sample size from N to $N + \delta - 1$ while preserving the mean $\bar{x}$.

We now further compare δ_M with $\bar{\delta}$ to indicate a preference for the sample reuse procedure. For $N = 2$, $\delta_M \equiv 2 > \bar{\delta} \equiv 1$. For $N = 3$, $\delta_M > \bar{\delta}$ for all t^2. For $N \geq 4$

$$\delta_M > \bar{\delta} \qquad \text{for } 0 \leq t^2 < \frac{N-1}{N-2}$$

$$\delta_M < \bar{\delta} \qquad \text{for } \frac{N-1}{N-2} \leq t^2 \leq a(N)$$

$$\delta_M > \bar{\delta} \qquad \text{for } t^2 \geq a(N)$$

where

$$a(N) = \frac{\left[12(N-1)^2(N-2) + (N-1)^4\right]^{1/2} - (N-1)^2}{2(N-2)}.$$

Further $a(N)$, bounded below by 2, increases monotonically to an upper bound of 3. It can also be shown that $N^{-1}t^2 \to 1$ as N grows. This implies that $\bar{\delta} \to 1$ as N increases. We also note that δ_M has a singularity at $t^2 = N - 1$, a value in which neighborhood t^2 is expected to be with a nonnegligible frequency. Hence δ_M will behave somewhat erratically with an appreciable frequency (Geisser, 1990a).

On the other hand it can be shown that the only time $\bar{\delta}$ can be very large is when t^2 is to the right of, but very close to, $(N-1)/(N-2)$, an interval which will be of exceedingly low probability for t^2 and hence for $\bar{\delta}$. Actually if $\bar{\delta}$ takes on a large value one might be suspicious of the exponential assumption for the sampling distribution.

In summary, we conclude that $\bar{\delta}$ is greatly preferred to δ_M and of course to $\hat{\delta}$ in this problem because of its stability and its drastically reduced relative influence on the effective sampling size appearing in the predictive distribution.

More generally in cases where not all of the hyperparameters appear in the first moment of the predictive distribution one can initially solve for those that do, say τ_1, of $\tau = (\tau_1, \tau_2)$ in this manner. Then one can consider a second or other moment that contains the remaining set τ_2 and possibly τ_1. Setting $\tau_1 = \bar{\tau}_1$ here from the first solution one then can obtain in a similar manner as before a solution for τ_2 (Geisser, 1990a).

The lesson to be relearned here is that when using data-driven methods to approximate hyperparameters, the actual result obtained in any particular case can be more compelling than the philosophical principle used to obtain the result.

Numerical illustration of Example 3.4. Previous experience with five randomly chosen light bulbs yields the following lifetimes in hours: 99.09, 1859.80, 111.90, 1899.82, 185.99. This results in $\bar{x} = 831.32$, $s^2 = 917413.9$, $\delta_M = 34.26$, $\bar{\delta} = 2.10$, and $\hat{\delta} = \infty$. We assume that the lifetime of bulbs is exponential and wish to predict the lifetime of a newly purchased bulb presumably ex-

Table 3.1 *Predictive Median for Varying δ*

δ	2.10	34.26	∞
Median	529.4	566.5	576.2

changeable with the previous ones. Now the expected lifetime irrespective of which δ is used is 831.22. We then compute the predictive median of lifetimes by setting

$$\left(1 + \frac{x_m}{(4+\delta)\bar{x}}\right)^{-(5+\delta)} = 0.5$$

and solving for x_m for the various values of δ corresponding to the methods (Table 3.1).

Although the predictive means are the same for the three methods, the predictive medians display considerable variation. Note that for $\gamma = \delta = 0$ from (3.1) the predictive mean is $1.25\bar{x} = 1039.15$. The predictive median is found by setting

$$\left(1 + \frac{x_m}{5\bar{x}}\right)^{-5} = 0.5$$

resulting in $x_m = 618.0$.

3.10 Prediction of a future fraction

Assume that $X^{(N+M)}$ is a random sample of copies from $F(x|\theta)$ so that

$$F\left(x_{(M)}|x^{(N)}\right) = \int F\left(x_{(M)}|\theta\right) dP\left(\theta|x^{(N)}\right)$$

and $X_{N+1}, \ldots, X_{N+M}$ are exchangeable random variables in the sense that all $M!$ permutations of $X_{(M)}$ have the same M-dimensional probability distribution. Now for some measurable set A let

$$Y_i = \begin{cases} 1 & \text{if } X_{N+i} \in A \\ 0 & \text{otherwise} \end{cases}$$

so that clearly the set $Y_1, Y_2, \ldots, Y_M$ is also exchangeable. Further let $R = \sum_{i=1}^{M} Y_i$ so that $M^{-1}R = \bar{Y}$ is the future fraction that will fall in A.

$$E(\bar{Y}) = \Pr[Y_i = 1] = \Pr(X_{N+i} \in A) = q$$

$$V(\bar{Y}) = \frac{1}{M^2} E\left[\sum_i (Y_i - q)\right]^2$$

$$= \frac{q(1-q)}{M} + \frac{(M-1)}{M} q(1-q)\rho$$

where for $i \neq j$

$$\rho = \frac{\text{cov}(Y_i, Y_j)}{\sqrt{V(Y_i)V(Y_j)}} = \frac{\Pr(Y_i = 1, Y_j = 1) - q^2}{q(1-q)}.$$

Since $Y_1, \ldots, Y_M$ are exchangeable the correlation ρ is the same for every i and j, $i \neq j$, and ρ is independent of M. Hence

$$\lim_{M \to \infty} V(\bar{Y}) = \rho q(1-q).$$

Now let

$$\tau = \tau_A(\theta) = \Pr[Y_i = 1|\theta] = \Pr[X_{N+i} \in A|\theta]$$

so that

$$\Pr[R = r|\theta] = \binom{M}{r} \tau^r (1-\tau)^{M-r}$$

$$\Pr[R = r] = \binom{M}{r} \int \tau^r (1-\tau)^{M-r} \, dP(\tau|x^{(M)})$$

where $P(\tau|x^{(N)})$ is calculated from $P(\theta|x^{(N)})$ via $\tau = \tau(\theta)$. Further it is clear that $\lim_{M \to \infty} M^{-1}R = \tau$.

There is a general theorem here due to de Finetti (1937) which is useful to cite.

Theorem. To every infinite sequence of exchangeable random binary variables $\{W_k\}$ corresponds a probability distribution $F_U(u)$

concentrated on $[0, 1]$ such that for every permutation of the integers $1, \ldots, M$ and any given integer $r \in [0, M]$

a. $\Pr[W_{j_1} = 1, \ldots, W_{j_r} = 1, W_{j_{r+1}} = 0, \ldots, W_{j_M} = 0]$

$$= \int_0^1 u^r (1 - u)^{M-r} \, dF(u)$$

b. $\Pr[\sum_{k=1}^M W_k = r] = \binom{M}{r} \int_0^1 u^r (1 - u)^{M-r} \, dF(u)$

c. $\lim_{M \to \infty} M^{-1} \sum_{k=1}^M W_k = U$

with U having distribution function $F(u)$.

For the simplest proof see Heath and Sudderth (1976). The theorem can be extended to show that every infinite sequence of exchangeable variables is a mixture of independent, identically distributed random variables, c.f. Loeve (1960).

Example 3.5. Let X_i be independent binary variables and $\Pr(X_i = 1) = \theta$, then $T = \sum X_i$ with probability

$$\binom{N}{t} \theta^t (1 - \theta)^{N-t}.$$

Suppose

$$p(\theta) = \frac{\Gamma(\alpha + \beta) \theta^{\alpha - 1} (1 - \theta)^{\beta - 1}}{\Gamma(\alpha) \Gamma(\beta)}$$

then

$$p(\theta | x^{(N)}) = \frac{\Gamma(N + \alpha + \beta) \theta^{t + \alpha - 1} (1 - \theta)^{N - t + \beta - 1}}{\Gamma(\alpha + t) \Gamma(N - t + \beta)}.$$

Hence for $R = \sum_{i=1}^M X_{N+i}$

$\Pr[R = r | t]$

$$= \int \binom{M}{r} \theta^r (1 - \theta)^{M-r} p(\theta | x^{(N)}) \, d\theta$$

$$= \frac{\Gamma(M + 1) \Gamma(N + \alpha + \beta) \Gamma(r + t + \alpha) \Gamma(M + N - r - t + \beta)}{\Gamma(r + 1) \Gamma(M - r + 1) \Gamma(\alpha + t) \Gamma(N - t + \beta) \Gamma(M + N + \alpha + \beta)}$$

Note that for $\alpha = \beta = 1$ or a uniform prior on θ the above reduces to

$$\Pr[R = r|t] = \frac{\binom{r+t}{t}\binom{M+N-t-r}{N-t}}{\binom{M+N+1}{M}}.$$

The latter result was also obtained without resort to a parametric formulation in example 3.2.

Example 3.6. Suppose $X^{(N)} = (X^{(d)}, X^{(N-d)})$ where $X^{(d)}$ represents copies fully observed from an exponential survival time density

$$f(x|\theta) = \theta e^{-\theta x}$$

and $X^{(N-d)}$ represents copies censored at $x_{d+1}, \ldots, x_N$, respectively. Hence

$$L(\theta) \propto \theta^d e^{-\theta N\bar{x}},$$

when $N\bar{x} = \Sigma_1^N x_i$. Suppose we are interested in the future number R of $X^{(M)}$ that will survive time y. Clearly,

$$\Pr[X > y|\theta] = e^{-\theta y} = \tau.$$

Further assume

$$p(\theta) = \frac{\gamma^\delta \theta^{\delta-1} e^{-\gamma\theta}}{\Gamma(\delta)}$$

then

$$p(\theta|x^{(N)}) = \frac{(\gamma + N\bar{x})^{d+\delta} \theta^{d+\delta-1} e^{-\theta(\gamma + N\bar{x})}}{\Gamma(d+\delta)}$$

and

$$\Pr[R = r|x^{(N)}, y] = \binom{M}{r} \int e^{-\theta y r}(1 - e^{-\theta y})^{M-r} p(\theta|x^{(N)})\, d\theta$$

$$= \binom{M}{r}(\gamma + N\bar{x})^{d+\delta} \sum_{j=0}^{M-r} \binom{M-r}{j}$$

$$\times (-1)^j[\gamma + N\bar{x} + y(r+j)]^{-(d+\delta)}.$$

If we let $\delta \to 0$, $\gamma \to 0$, then

$$\Pr\left[R = r|x^{(N)}, y\right]$$
$$= \binom{M}{r}(N\bar{x})^d \sum_{j=0}^{M-r} \binom{M-r}{j}(-1)^j[N\bar{x} + y(r+j)]^{-d}$$

the so-called noninformative case, Geisser (1982).

Numerical illustration of Example 3.6. A random sample of size $N = 20$ from an exponential distribution with mean 1 was censored at $x = 2$ yielded $\bar{x} = 0.749$ and $d = 18$. In Figure 3.1, a plot of $\Pr[R = r|x^{(20)}, y]$ for $M = 10$, $y = \log 2$ is superimposed on the true binomial distribution $B(10, 0.5)$. In this case the predictive probability function tends to approximate the true probabilities except that as expected it tends to be more diffuse.

Example 3.7. We suppose that now,

$$f(x|\theta) = \theta e^{-\theta(x-\gamma)}, \qquad \theta > 0, \qquad x > \gamma > -\infty$$

the two parameter exponential density (Geisser, 1984, 1985b). Now let $m_d = \min(x_1, \ldots, x_d)$ and order the censored values as follows: $x_{(1)} \geq x_{(2)} \geq \cdots \geq x_{(N-d)}$. We shall also assume that the smallest value m is one of the fully observed values, $m = m_d \leq x_{(N-d)}$. This will often happen in practice and avoids much heavier computation. Letting $p(\theta, \gamma) \propto \theta^{-1}$, we obtain the posterior distributions,

$$p(\gamma|\theta, x^{(N)}) = N\theta e^{\theta N(\gamma - m)} \qquad \text{for } \gamma < m$$
$$p(\theta|x^{(N)}) = [N(\bar{x} - m)]^{d-1} \theta^{d-2} e^{-\theta N(x-m)} \qquad \theta > 0.$$

From the above we can calculate the predictive distribution function or survival function of a future observation Z. The latter result is easily verified to be

$$\Pr(Z > z) = \begin{cases} \dfrac{N^d(\bar{x} - m)^{d-1}}{(N+1)[z - m + N(\bar{x} - m)]^{d-1}}, & z > m, \\[4mm] 1 - (N+1)^{-1}\left(\dfrac{\bar{x} - m}{\bar{x} - z}\right)^{d-1}, & z \leq m. \end{cases}$$

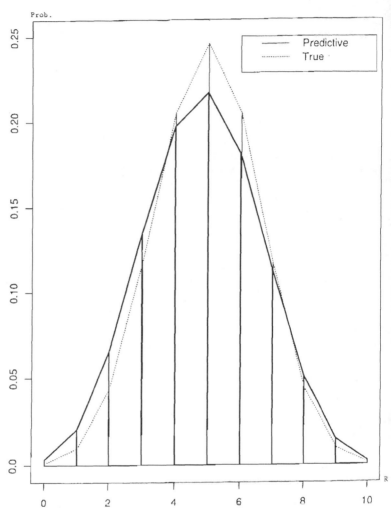

Figure 3.1   *Predictive probabilities superimposed on true binomial proba-bilities.*

Consider now the fraction of a set of future values $Z_1, \ldots, Z_M$ that exceeds a given value, z say. Let

$$Y_i = \begin{cases} 1 & \text{if } Z_i > z, \\ 0 & \text{if } Z_i \leq z \end{cases}$$

for $i = 1, \ldots, M$ so that $\bar{Y} = M^{-1}[Y_1 + \cdots + Y_M]$ is the required fraction. Now we have the conditional probability

$$\Pr[Z_i > z | \theta, \gamma] = \tau = \min(e^{-\theta(z-\gamma)}, 1),$$

so that the probability function of $\bar{Y}$ can be obtained from

$$\Pr\left[\bar{Y} = \frac{r}{M} \bigg| z\right] = \int \binom{M}{r} \tau^r (1-\tau)^{M-r} \, dP(\tau|x),$$

where $P(\tau|x)$ is the posterior distribution of the random variable τ. The distribution of τ conditional on θ is easily calculated to be

$$P(\tau|\theta, x^{(N)}) = \begin{cases} 0, & \tau \leq 0. \\ \tau^N e^{\theta N(z-m)}, & 0 < \tau < \min(e^{\theta(m-z)}, 1), \\ 1, & \tau \geq 1, \end{cases}$$

with

$$\Pr[\tau = 1 | \theta, x^{(N)}] = \begin{cases} 1 - e^{\theta N(z-m)} & \text{if } z < m, \\ 0 & \text{if } z \geq m. \end{cases}$$

By taking the expectation of the conditional distribution (density) of τ with respect to the posterior distribution of θ, the unconditional posterior distribution of τ is obtained. For $0 < \tau < 1$ and

$$I_a(p) = \int_0^p \frac{e^{-u} u^{a-1}}{\Gamma(a)} \, du,$$

we obtain

$$P(\tau|x^{(N)})$$

$$= \begin{cases} \tau^N \left(\dfrac{\bar{x} - m}{\bar{x} - z} \right)^{d-1}, & z \leq m, \\[3mm] \tau^N \left(\dfrac{\bar{x} - m}{\bar{x} - z} \right)^{d-1} I_{d-1}\left[N \left(\dfrac{\bar{x} - z}{m - z} \right) \log \tau \right] \\[3mm] \quad + 1 - I_{d-1}\left[N \left(\dfrac{\bar{x} - m}{m - z} \right) \log \tau \right], & m < z \neq \bar{x}, \\[3mm] \dfrac{N^{d-1} \tau^N [-\log \tau]^{d-1}}{\Gamma(d)} + 1 - I_{d-1}(-N \log \tau), & z = \bar{x}, \end{cases}$$

and

$$\Pr(\tau = 1|x) = \begin{cases} 1 - \left(\dfrac{\bar{x} - m}{\bar{x} - z} \right)^{d-1}, & z < m, \\[3mm] 0, & z \geq m. \end{cases}$$

We then obtain

$$\Pr\left(\bar{Y} = \frac{r}{M} \middle| z \right)$$

$$= \begin{cases} \left(\dfrac{\bar{x} - m}{\bar{x} - z} \right)^{d-1} \left(\dbinom{N + r - 1}{r} \right) \Big/ \left(\dbinom{N + M}{M} \right), & r < M, \quad z < m, \\[4mm] 1 - \dfrac{M}{N + M} \left(\dfrac{\bar{x} - m}{\bar{x} - z} \right)^{d-1}, & r = M, \quad z < m, \\[4mm] N \dbinom{M}{r} \sum_{j=0}^{M-r} \dbinom{M - r}{j} \\[3mm] \quad \times \dfrac{(-1)^j}{(N + r + j)} \left(1 + \dfrac{(r + j)(z - m)}{N(\bar{x} - m)} \right)^{-(d-1)}, & m < z, \\[4mm] \dbinom{M}{r} \sum_{j=0}^{M-r} \dbinom{M - r}{j} (-1)^j \left(1 + \dfrac{r + j}{N} \right)^{-d}, & z = \bar{x}. \end{cases}$$

and note that from de Finetti's theorem

$$\lim_{M \to \infty} \bar{Y} = \tau$$

with distribution function $P(\tau | x^{(N)})$.

If we assume the conjugate prior $p(\gamma, \theta) = p(\gamma | \theta) p(\theta)$ such that

$$p(\gamma | \theta) = N_0 \theta e^{\theta N_0 (\gamma - m_0)}, \qquad \gamma < m_0$$

$$p(\theta) = \left[N_0 (\bar{x}_0 - m_0) \right]^{d_0 - 1} \theta^{d_0 - 2} e^{-\theta N_0 (\bar{x}_0 - m_0)} \Big/ \Gamma(d_0 - 1)$$

$\theta > 0$, $\bar{x}_0 > m_0$ and $1 < d_0 \le N_0$, then it is easy to verify that the results are such that for d, N, $\bar{x}$, and m we just substitute $d^* = d_0 + d$, $N^* = N_0 + N$, $\bar{x}^* = (N_0 + N)^{-1}(N_0 \bar{x}_0 + N\bar{x})$ and $m^* = \min(m_0, m)$ in the formulas.

Numerical illustration of Example 3.7. Grubbs (1971) obtained data on the mileage at time of failure in service of 19 personnel carriers. The ordered values were

162, 200, 271, 302, 393, 508, 539, 629, 706, 777
884, 1008, 1101, 1182, 1463, 1603, 1984, 2355, 2880.

He assumes that a critical mission mileage of 200 miles is ordinarily needed. We compute the chance that the next carrier will exceed 200 miles, and that 16 or more out of the next 20 will exceed 200 miles. Finally we calculate the chance that in a very large number of carriers, at least 80% of them will exceed 200 miles. We obtain, for $N = d = 19$, $m = 162$ and $\bar{x} = 997.21$, that

$$\Pr[Z \ge 200] = 0.910$$

$$\Pr\left[\bar{Y}_{20} \ge \frac{16}{20} \Big| z = 200 \right] = 0.938$$

$$\Pr[\tau \ge 0.8 | z = 200, \bar{x}, m] = 0.967.$$

Example 3.8. Suppose now that $X_i \sim N(\mu, \sigma^2)$ $i = 1, \ldots, N + M$. Let $p(\mu, \sigma^2) \propto 1/\sigma^2$. Then it is easy to show that

$$p(\mu, \sigma^2 | x^{(N)}) = p(\mu | \sigma^2, x^{(N)}) p(\sigma^2 | x^{(N)})$$

where $\mu | \sigma^2 \sim N(\bar{x}, \sigma^2/N)$ and $(N - 1)s^2/\sigma^2 \sim \chi^2_{N-1}$.

We note that using the Jeffrey's invariance criterion directly would have led to

$$p(\mu, \sigma^2) \propto |I(\mu, \sigma^2)|^{1/2} \propto \frac{1}{\sigma^3}.$$

It is more usual to treat the invariance for μ and σ^2 separately so that for known σ^2,

$$p(\mu) \propto |I(\mu)|^{1/2} \propto \text{constant}$$

and

$$p(\sigma^2) \propto |I(\sigma^2)|^{1/2} \propto \frac{1}{\sigma^2}.$$

Thus assuming a priori independence

$$p(\mu, \sigma^2) \propto p(\mu) p(\sigma^2) \propto \frac{1}{\sigma^2}.$$

This prior could be obtained directly using the Kullback loss function at the end of Section 6, by demonstrating that it leads to the optimal predictive density of X_{N+1} among all such predictive densities that are restricted to the form given in Example 3.7. This results in $X_{N+1} \sim S[N-1, \bar{x}, s^2(N^2-1)/N]$ or that

$$(X_{N+1} - \bar{x})\sqrt{N}/s\sqrt{N+1} = t_{N-1}.$$

Further, for R the number of $X_{N+1}, \ldots, X_{N+M}$ that exceeds x

$$\Pr[R = r|x^{(N)}] = \binom{M}{r} \int_0^\infty \int_{-\infty}^\infty \theta^r (1-\theta)^{M-r} p(\mu, \sigma^2|x^{(N)}) \, d\mu \, d\sigma^2$$

where the random variable

$$\theta = P[X > x|\mu, \sigma^2] = 1 - \Phi\left(\frac{x - \mu}{\sigma}\right)$$

is a function of the random variables μ and σ. This integral, as well as

$$\Pr[\,R \geq r|x^{(N)}] = \sum_{j=r}^{M} P(R = j|x^{(N)}),$$

can be computed numerically. For large M we know that

$$\lim_{M \to \infty} \frac{R}{M} = \theta$$

by de Finetti's theorem. Further let $\beta = (\mu - x)/\sigma$, then

$$F_\theta(\theta_0|x^{(N)}) = \Pr[\beta \leq \Phi^{-1}(\theta_0)] = F_\beta[\Phi^{-1}(\theta_0)]$$

the distribution function of β, where the density is given by

$$f(\beta) = \frac{\sqrt{N}\,e^{-N\beta^2/2}}{\sqrt{2\pi}\,\Gamma\!\left(\dfrac{N-1}{2}\right)} \sum_{j=0}^{\infty} \frac{\left(\sqrt{2}\,\beta Nd\right)^j \Gamma((N+j-1)/2)}{j!(1+Nd^2)^{(N+j-1)/2}}$$

for

$$d = \frac{\bar{x}_N - x}{(N-1)^{1/2}s}$$

and $(N-1)s^2 = \sum_1^N (x_i - \bar{x})^2$ (Geisser, 1967, 1987a). One can also show that

$$F_\theta(\theta_0|x^{(N)}) = \Pr\!\left[T \leq d\sqrt{N(N-1)}\,\right]$$

where T is the noncentral student variate with $N-1$ degrees of freedom and noncentrality parameter $\lambda = \sqrt{N}\,\Phi^{-1}(1 - \theta_0)$ (Aitchison 1964). Now since $\Phi(\beta) = \theta$,

$$\Pr(R = r|x^{(N)}) = \binom{M}{r} \int_{-\infty}^{\infty} \Phi'(\beta)[1 - \Phi(\beta)]^{M-r} f(\beta)\,d\beta$$

which can be evaluated numerically. It can be shown that as N

grows the distribution of β tends to $N(\bar{\beta}, \sigma_\beta^2)$ where

$$\bar{\beta} = k\left(\frac{2N-3}{2N-2}\right)^{1/2}$$

$$\sigma_\beta^2 = N^{-1} + \frac{k^2}{2(N-1)},$$

and $k = d\sqrt{N-1}$ (Geisser, 1967). Therefore

$$F_\theta\left(\theta_0 | x^{(N)}\right) \doteq 1 - \Phi\left[\frac{\bar{\beta} + \Phi^{-1}(1-\theta_0)}{\sigma_\beta}\right]$$

and

$$\Pr\left(R = r | x^{(N)}\right) \doteq \binom{M}{r}\int_{-\infty}^{\infty} \Phi^r(\beta)[1 - \Phi(\beta)]^{M-r} n(\beta)\, d\beta$$

where $n(\beta)$ is the density of a normal variate with mean $\bar{\beta}$ and variance σ_β^2. All of these results will also hold more generally if the normal variables $X_1, \ldots, X_{N+M}$ are exchangeable with common correlation $\rho \geq 0$.

Numerical illustration of Example 3.8. A user of expensive ball bearings is planning to purchase 1000 whose advertised diameter is 0.8 (mm). A random sample of 30 yields a mean diameter (mm), $\bar{x} = 0.8056$ with $s = 0.00928$. For a certain industrial use it is necessary that the diameter (mm) of a ball bearing exceed $z = 0.79$ for its efficient use. The user will be satisfied with his purchase if he has a degree of assurance that at least 90% of his purchase can be efficiently used. We now compute the probability, that at least 900 out of the 1000 will meet the need for efficient use.

Now, from the noncentral T distribution or from the distribution of β we calculate the exact asymptotic value, $\Pr(\theta \geq 0.9 | 0.79) = 0.913$. We can calculate

$$\sum_{r-900}^{1000} \Pr[R = r | x^{(N)}]$$

numerically with the use of a computer. This turns out to be 0.910. However, if $M = 20$ so that 90% or more yielded $R \geq 18$, the "exact" calculation yields 0.881. Hence the asymptotic value is 3.6% larger. For $M = 100$ and $R \geq 90$ we obtain 0.894 and the error is less than 1.8%. Hence depending on M and the tolerable error, the easier asymptotic calculation may serve the intended purposes. Calculations of this sort are only available in a Bayesian approach.

3.11 Linear regression

Let the observables Y be related as

$$Y = X\beta + U$$

$$Y = \begin{pmatrix} Y_1 \\ \vdots \\ Y_n \end{pmatrix}, \qquad X = \begin{pmatrix} x_{11} & ,\ldots, & x_{1p} \\ \vdots & & \vdots \\ x_{N1} & ,\ldots, & x_{Np} \end{pmatrix}, \qquad \beta = \begin{pmatrix} \beta_1 \\ \vdots \\ \beta_p \end{pmatrix}$$

where X is known, β unknown, and

$$U = \begin{pmatrix} U_1 \\ \vdots \\ U_N \end{pmatrix} \sim N(0, \sigma^2 I_N).$$

Hence,

$$f_Y(y) = \frac{e^{-(1/2\sigma^2)(y - X\beta)'(y - X\beta)}}{(2\pi)^{N/2} \sigma^N}$$

where y is the realization of the vector Y.
   For the noninformative prior

$$p(\beta, \sigma^2) \propto \frac{1}{\sigma^2},$$

$$p(\beta, \sigma^2 | y) \propto \frac{1}{(\sigma^2)^{(N/2)+1}} e^{-(1/2\sigma^2)(y - X\beta)'(y - X\beta)}.$$

For

$$\hat{\beta} = (X'X)^{-1}X'y$$

we have the identity

$$(y - X\beta)'(y - X\beta) = (y - X\hat{\beta})'(y - X\hat{\beta}) + (\beta - \hat{\beta})'X'X(\beta - \hat{\beta}).$$

Now by integrating out σ^2, where $(N - p)s^2 = (y - X\hat{\beta})'(y - X\hat{\beta})$,

$$p(\beta|y) = \frac{\Gamma(N/2)}{\pi^{p/2}\Gamma((N-p)/2)\left|s^2(X'X)^{-1}\right|^{1/2}(N-p)^{p/2}}$$

$$\times \left[1 + \frac{(\beta - \hat{\beta})'X'X(\beta - \hat{\beta})}{(N-p)s^2}\right]^{-N/2}$$

or

$$\beta \sim S_p\left[N - p, \hat{\beta}, (N - p)s^2(X'X)^{-1}\right]$$

a multivariate student distribution.

Suppose we are interested in predicting a set of M new variates

$$Z = \begin{pmatrix} Z_1 \\ \vdots \\ Z_M \end{pmatrix}$$

at known design matrix

$$W = \begin{pmatrix} w_{11}, \ldots, w_{1p} \\ \vdots \qquad \vdots \\ w_{M1}, \ldots, w_{Mp} \end{pmatrix}$$

such that

$$Z = W\beta + U$$

where

$$U \sim N(0, \sigma^2 I_M).$$

Then for the future set Z

$$f(z|y) = \int f(z|\sigma^2, \beta, W) p(\sigma^2, \beta | y, X) \, d\sigma^2 \, d\beta.$$

Let

$$A = I + W(X'X)^{-1}W'$$

then we can calculate

$$f(z|y) = \frac{\Gamma((N+M-p)/2)}{\pi^{M/2}\Gamma((N-p)/2)(N-p)^{M/2}|s^2 A|^{1/2}}$$

$$\times \left[1 + \frac{(z-W\hat{\beta})' A^{-1}(z-W\hat{\beta})}{(N-p)s^2}\right]^{-(N+M-p)/2}$$

or

$$Z|y \sim S_M\left[N-p, W\hat{\beta}, (N-p)s^2 A\right].$$

A joint predictive region for Z can be obtained from $Z|y$ by noting that

$$\frac{(Z-W\hat{\beta})' A^{-1}(Z-W\hat{\beta})}{(N-p)s^2} = F_{M, N-p}$$

an F variate with M and $N-p$ degrees of freedom. For the special case $M = p = 1$ so that we are dealing with i.i.d. normal variates, the predictive density is easily obtained as

$$\frac{(Z-\bar{y})\sqrt{N}}{s\sqrt{1+N}} \sim t_{N-1}.$$

3.12 Low structure Bayesian prediction

Let $X_1, \ldots, X_N, X_{N+1}, \ldots, X_{N+M}$ be exchangeable and specify that conditional on the realized values of $X_i = x_i$, $i = 1, \ldots, N$ the predictive probability for a given R, the number of future observables $X_{N+1}, \ldots, X_{N+M}$, that will be observed between the consecutive observed order statistics $y_0 < y_1 < \cdots < y_N < y_{N+1}$ where $y_0 = -\infty$ and $y_{N+1} = \infty$, will be equal (Hill, 1968). Although this probability will differ for each $R = 0, 1, \ldots, M$, it is assumed constant for any given R. For $M = 1$, this assumption was proposed by Hill (1968) and the assumption here is easily derivable from Hill's. This implies that the results given previously in Section 1.2 or Chapter 2 will coincide with these. The reasonableness of this assumption as a Bayesian evaluation for all N was foundationally demonstrated by Lane and Sudderth (1984). The difference between the approach here and in Chapter 2 is mainly in the interpretation of a conditional predictive probability and one based on repeating sampling. This has been extended to right censored observations and ties by Berliner and Hill (1988).

References

Aitchison, J. (1964). Bayesian tolerance regions. *Journal of the Royal Statistical Society B* **26**, 161–175.

Akaike, H. (1978). A new look at the Bayes procedure. *Biometrika* **65**, 53–59.

Barnard, G. A. (1954). Sampling inspection and statistical decisions. *Journal of the Royal Statistical Society B* **16**, 151–174.

Bayes, T. (1763). An essay towards solving a problem in the doctrine of chances. *Philosophical Transactions of the Royal Society of London* **53**, 370–418.

Berliner, L. M., and Hill, B. M. (1988). Bayesian nonparametric survival analysis. *Journal of the American Statistical Association* **83**, 772–784.

Bernardo, J. M. (1979). Reference posterior distributions for Bayesian inference (with discussion). *Journal of the Royal Statistical Society B* **41**, 113–147.

Box, G. E. P. and Tiao, G. C. (1973). *Bayesian Inference in Statistical Analysis* Reading, MA: Addison-Wesley.

Cornfield, J. (1969). The Bayesian outlook and its applications. *Biometrics* **25**(4), 617–657.

de Finetti, B. (1937). Le Prevision: ses lois logiques, ses sources subjectives. *Ann. Inst. Poincare*, tome VIII, fasc. 1, 1–68. Reprinted in *Studies in Subjective Probability*. Melbourne, FL: Krieger, 1980 (English translation).

De Groot, M. H. (1970). *Optimal Statistical Decisions*. New York: McGraw-Hill.

Freedman, D. A., and Purves, R. A. (1969). Bayes' method for bookies. *Annals of Mathematical Statistics* **40**, 1177–1186.

Geisser, S. (1967). Estimation associated with linear discriminants. *Annals of Mathematical Statistics* **38**, 807–817.

Geisser, S. (1979). "Discussion" of Bernardo, "Reference posterior distributions for Bayesian inference," *Journal of the Royal Statistical Society B* **41**, 136–137.

Geisser, S. (1980b). Predictive sample reuse for censored data. In *Bayesian Statistics*, J. M. Bernardo, et al. (ed.). Valencia, Spain: University Press.

Geisser, S. (1982). Aspects of the predictive and estimative approaches in the determination of probabilities. *Biometrics* **38**, suppl. 75–93.

Geisser, S. (1984). Predicting Pareto and exponential observables. *Canadian Journal of Statistics* **12**, 143–152.

Geisser, S. (1985a). On the predicting of observables: A selective update. In *Bayesian Statistics 2*, J. M. Bernardo et al. (eds.). Amsterdam: North-Holland, 203–230.

Geisser, S. (1985b). Interval prediction for Pareto and Exponential observables. *Journal of Econometrics* **29**, 173–185.

Geisser, S. (1987a). Some remarks on exchangeable normal variables with applications. In *Contributions to the Theory and Application of Statistics*, A. Gelfand (ed.), New York: Academic Press, 127–153.

Geisser, S. (1990a). On hierarchical Bayes procedures for predicting simple exponential survival. *Biometrics* **46**, 225–230.

Grubbs, F. E. (1971). Fiducial bounds on reliability for the two-parameter negative exponential distribution. *Technometrics* **13**, 873–876.

Heath, D., and Sudderth, W. (1976). De Finetti's theorem on exchangeable variables. *The American Statistician* **30**(4), 188–189.

Hill, B. M. (1968). Posterior distribution of percentiles: Bayes theorem from a finite population. *Journal of the American Statistical Association* **63**, 677–691.

Jeffreys, Harold (1946). An invariant form for the prior probability in estimation problems. *Proceedings of the Royal Society of London A* **186**, 453–454.

Lane, D. A., and Sudderth, W. D. (1984). Coherent predictive inference. *Sankhyā: The Indian Journal of Statistics*, **46**(A, 2), 166–185.

Loeve, M. (1960). *Probability Theory*, New York: van Nostrand.

Morris, C. N. (1983). Parametric empirical Bayes inference: theory and applications. *Journal of the American Statistical Association.* **78**, 47–55.

Raiffa, H., and Schlaifer, R. (1961). *Applied Statistical Decision Theory.* Cambridge: Harvard University Press.

Stigler, S. M. (1982). Thomas Bayes and Bayesian inference. *Journal of the Royal Statistical Society A* **145**(2), 250–258.

Zellner, A. (1977). Maximal data information prior distributions. In *New Developments in the Application of Bayesian Methods*, A. Aykac and C. Brumat (eds.). Amsterdam: North-Holland.

CHAPTER 4

Selecting a statistical model and predicting

Statistical models are often simplifications of extremely complicated phenomena and it is a mistake to necessarily assume that any such model is a true representation of the underlying process. What is assumed is that the model is useful for some purpose such as a parsimonious description or predicting future occurrences. Hence selecting a model, irrespective of its truth, that best serves such purposes is the goal of model selection.

In this chapter we first start with a situation where several models are entertained and indicate how one selects the most probable model and how this may differ from a pure prediction problem. From there we present several large sample model selection procedures as well as several useful sample reuse procedures when strong prior assumptions are muted. Further, once a model is entertained, methods for criticizing it are presented. Predictive methods for the detection of discordant observations and diagnostics for highly influential observations are also presented.

4.1 Complete Bayesian model selection

Suppose there are K possible models $M_1, \ldots, M_K$ for consideration with regard to a data set $x^{(N)}$. Let

$$\Pr(M_k) = q_k, \qquad k = 1, \ldots, K \quad \text{and} \quad \sum q_k = 1.$$

Assume that for M_k the model specification for the data is

$$f\left(x^{(N)}, \theta_k | M_k\right) = f\left(x^{(N)} | \theta_k, M_k\right) p\left(\theta_k | M_k\right)$$

where θ_k represents a set of parameters specified by M_k.

Now for any M_k we can also calculate

$$p\left(\theta_k | x^{(N)}, M_k\right) \propto f\left(x^{(N)} | \theta_k, M_k\right) p\left(\theta_k | M_k\right).$$

Hence to predict a future value X_{N+1} given M_k

$$f\left(x_{N+1} | x^{(N)}, M_k\right) = \int f\left(x_{N+1} | x^{(N)}, \theta_k, M_k\right) p\left(\theta_k | x^{(N)}, M_k\right) d\theta_k.$$

Now if we want to select the single most probable model of the several entertained we would select M_{k*}, such that

$$\max_k \Pr\left[M_k | x^{(N)}\right] = \Pr\left[M_{k*} | x^{(N)}\right],$$

where

$$\Pr\left[M_k | x^{(N)}\right] = \frac{q_k f\left(x^{(N)} | M_k\right)}{\sum\limits_k q_k f\left(x^{(N)} | M_k\right)} = q_k',$$

and

$$f\left(x^{(N)} | M_k\right) = \int f\left(x^{(N)}, \theta_k | M_k\right) d\theta_k.$$

However, using M_{k*} for predicting would not be optimal unless there was a sufficient penalty for using more than one model for prediction. This is easily seen since the unconditional predictive density is

$$f\left(x_{N+1} | x^{(N)}\right) = \sum_{k=1}^{K} q_k' f\left(x_{N+1} | x^{(N)}, M_k\right).$$

Hence optimal prediction would in general apply a loss function to this mixture of densities.

Example 4.1. Suppose we have a set of data with two labels such that the realized values of

$$X^{(N)} = x^{(N)} = \left(x^{(N_1)}, x^{(N_2)} \right), \qquad N_1 + N_2 = N$$

$$X^{(N_1)} = x^{(N_1)} = \left(x_1, \ldots, x_{N_1} \right), \; X^{(N_2)} = x^{(N_2)} = \left(x_{N_1+1}, \ldots, x_{N_1+N_2} \right).$$

Assume the model specifications are

M_1: X_i $i = 1, \ldots, N_1 + N_2$ are i.i.d. with density

$$f(x|\theta) = \theta e^{-\theta x}$$

$$p(\theta|M_1) \propto \theta^{\delta-1} e^{-\gamma \theta}.$$

M_2: X_i, $i = 1, \ldots, N_1$ are i.i.d. with density

$$f(x|\theta) = \theta_1 e^{-\theta_1 x}$$

$$p(\theta_1|\delta_1, \gamma_1) \propto \theta_1^{\delta_1-1} e^{-\gamma_1 \theta_1}$$

independent of

X_i, $i = N_1 + 1, \ldots, N_1 + N_2$ that are i.i.d. with density

$$f(x|\theta) = \theta_2 e^{-\theta_2 x}$$

$$p(\theta_2|\gamma_2, \delta_2) \propto \theta_2^{\delta_2-1} e^{-\gamma_2 \theta_2}.$$

For prediction conditional on M_1,

$$f\left(z_j|x^{(N)}, M_1\right) = \frac{(N + \delta)(N\bar{x} + \gamma)^{N+\delta}}{\left(N\bar{x} + \gamma + z_j\right)^{N+\delta+1}}$$

where $j = 1, 2$ represents the label. Here the label is irrelevant. Further, conditional on M_2,

$$f\left(z_j|x^{(N_j)}, M_2\right) = \frac{(N_j + \delta_j)(N_j\bar{x}_j + \gamma_j)^{N_j+\delta_j}}{\left(N_j\bar{x}_j + \gamma_j + z_j\right)^{N_j+\delta_j+1}},$$

where $\bar{x}_j$ represents the sample mean from the jth label. For selection of the "best" model choose the larger of

$$q_1 f\left(x^{(N)}|M_1\right) = q_1 \Gamma(N+\delta)\gamma^\delta/\Gamma(\delta)[N\bar{x}+\gamma]^{N+\delta} \alpha\, q_1'$$

or

$$q_2 f\left(x^{(N)}|M_2\right) = q_2 f\left(x^{(N_1)}|M_2\right) f\left(x^{(N_2)}|M_2\right) \alpha\, q_2'$$

where

$$f\left(x^{(N_j)}|M_2\right) = \Gamma(N_j+\delta_j)\gamma_j^{\delta_j}/\Gamma(\delta_j)\left[N_j\bar{x}_j+\gamma_j\right]^{N_j+\delta_j}.$$

Note that unless $f(x^{(N)}|M_k)$ is a proper density there may be some difficulty in the implementation of this procedure.

The question then is what do we do once we select the best model? Presumably it affords the "best" single description among those entertained. Should we now use it for prediction? If we do, we know that it is not optimal under any loss function except one that reflects a principle of parsimony that states that one should only use one of the K models for prediction. Many scientific workers appear to agree with this when the K models are physically meaningful and exhaustive. In usual circumstances, however, optimal statistical prediction depends on a calculation of

$$q_1' f\left(z_j|x^{(N)}, M_1\right) + q_2' f\left(z_j|x^{(N)}, M_2\right) = f\left(z_j|x^{(N)}\right),$$

and applying a loss function to this density.

4.2 Large sample model selection

Often the specification of a proper prior for θ is either too difficult or not terribly meaningful as these models are at best only adequate approximations to data-generating mechanisms. One asymptotic approach is as follows: Suppose for nested models the number of unknown parameters of M_k is m_k, then Schwarz (1978) shows, for the exponential class of likelihoods and a special class of priors that puts positive probability on the subspaces that correspond to the competing models, that the large sample limit of the Bayes procedure depends on the first two terms of the following expression,

$$\log q_k f\left(x^{(N)}|M_k\right) \doteq \log f\left(x^{(N)}|\hat{\theta}_k, M_k\right) - m_k \log \sqrt{N}$$
$$+ \frac{1}{2}m_k \log \frac{\pi}{\lambda} + \log q_k,$$

where $\hat{\theta}_k$ is the MLE (maximum likelihood estimator) and $\lambda > 0$. Further the last two terms on the right-hand side are bounded in N for any given $x^{(N)}$ and hence are small relative to the rest under the assumption of an equal fixed loss (or a loss that is never outside of fixed bounds) for choosing the wrong model. This tends to be independent of the q_k's as well if they are bounded away from 0 and 1. If the q_k's are equal, the approximation is better. Here we use the first two terms and choose the model that has the largest $\log S_k$ where

$$\log S_k = \log f\left(x^{(N)}|\hat{\theta}_k, M_k\right) - m_k \log \sqrt{N}.$$

Leonard (1982) has pointed out that the proof by Schwarz will hold for nonnested situations as well.

Another approach was proposed by Akaike (1973). This involves the quantity

$$\log f\left(x^{(N)}|\hat{\theta}_k, M_k\right) - m_k + O\left(\frac{1}{N}\right),$$

derivable from an information theoretic approach. One chooses the largest $\log A_k$ using the first two terms, i.e.,

$$\log A_k = \log f\left(x^{(N)}|\hat{\theta}_k, M_k\right) - m_k$$

to select the model. Note that in situations where both procedures apply,

$$\log \frac{A_k}{S_k} = m_k\left[\log \sqrt{N} - 1\right].$$

This difference grows with N.

Also we can express

$$A_k = \left(\sqrt{N}/e\right)^{m_k} S_k.$$

For nested situations, where

$$m_{k'} - m_k = d,$$

selecting M_k in preference to $M_{k'}$, requires

$$A_k > A_{k'}$$

or

$$S_k > \left(\sqrt{N}/e\right)^d S_{k'}.$$

The relative penalty (the coefficients of $S_{k'}$) increases rapidly with N for $N \geq 8$. Hence the Akaike procedure is also an asymptotic Bayes procedure but with a severe relative penalty incurred when a false lower dimensional model is selected as opposed to selecting a false higher dimensional model. For prediction problems this often makes good sense.

Other penalized likelihood methods are generally of the form

$$\log f\left(x^{(N)}|\hat{\theta}_k, M_k\right) - m_k C$$

where C may be a constant, a function of N or even of $x^{(N)}$ (Smith and Spiegelhalter, 1980; Atkinson, 1980; San Martini and Spezzaferri, 1984).

4.3 Sample reuse approaches to model selection

Let $X_1, \ldots, X_N$ be a random sample from $f(x|\theta_k, M_k)$. When $f_k = f(x_1, \ldots, x_N|M_k)$ does not exist but $f(x_j|x_j^{(N-1)}, M_k)$ does, as will occur with an improper prior for θ_k, let

$$\mathscr{L}_k = \prod_{j=1}^{N} \hat{f}\left(x_j|x_j^{(N-1)}, M_k\right),$$

where $\hat{f}$ is an "estimate" of $f(x|\theta_k, M_k)$. Now $\hat{f}$ can be either the maximum likelihood estimate, i.e., $\hat{f} = f(x_j|\hat{\theta}_{k(j)}, M_k)$ where $\hat{\theta}_{k(j)}$ is the MLE with x_j deleted, or some $\hat{f}$ based on a loss function, say, the Kullback divergence or the conditional predictive density based on an improper prior density. The rule is to choose $\mathrm{Max}_k \, \mathscr{L}_k = \mathscr{L}_{k*}$ so that the best choice is M_{k*} (Geisser and Eddy, 1979). Similar considerations involve

$$P_k = \prod_{j=t}^{N} \hat{f}\left(x_j|x^{(j-1)}, M_k\right)$$

where

$$\hat{f}\left(x_j|x^{(j-1)}, M_k\right) = \hat{f}\left(x_j|x_{j-1}, \ldots, x_1, M_k\right).$$

However, we start at the first t for which in all M_k

$$f_k\left(x_t|x_{t-1}, \ldots, x_1, M_k\right)$$

exists noting that in the Bayesian case the prior was such that $f(x_1, \ldots, x_N|M_k)$ did not exist for all M_k. Hence we use the product of existing estimative or conditional predictive densities

$$P_k = \prod_{j=t}^{N} \hat{f}\left(x_j|x_j^{(N-1)}, M_k\right)$$

and choose M_{k*} such that

$$\mathrm{Max}_k \, P_k = P_{k*}.$$

When t is sequenced in time, there is a natural order to this procedure. When this is not the case the value for P_k will depend on which of the xs were designated as $x_1, \ldots, x_{t-1}$. For very large sample sizes it may not matter very much (cf. Dawid, 1984). For moderate sample sizes one could repeat the process using all possible subsets of the initial t observations drawn from the N and comparing the averaged values of the various $\bar{P}_1, \ldots, \bar{P}_K$ for purposes of selection. Also in certain situations another type of adjustment can be made to k as described in the latter part of the following example.

Example 4.2. Suppose we have two sets of data and we want to select either

$$M_1: X^{(N)} = (X^{(N_1)}, X^{(N_2)}) \text{ i.i.d. with density}$$

$$f(x|\theta) = \theta e^{-\theta x}$$

or

$$M_2: X_j \quad j = 1, \ldots, N_1$$

with density

$$f(x|\theta_1) = \theta_1 e^{-\theta_1 x}$$

and for $j = N_1 + 1, \ldots, N_1 + N_2$, the density is

$$f(x|\theta_2) = \theta_2 e^{-\theta_2 x}.$$

Hence under M_1 the approach that uses the MLE is

$$f\left(x_j|\hat{\theta}_{k(j)}, M_1\right) = \frac{1}{\bar{x}_{(j)}} e^{-x_j / \bar{x}_{(j)}}$$

where $(N - 1)\bar{x}_{(j)} = N_1 \bar{x}_1 + N_2 \bar{x}_2 - x_j$. The conditional predictive density approach yields

$$\bar{f}(x_j|x_{(j)}, M_1) = (N - 1)\left[(N - 1)\bar{x}_{(j)}\right]^{N-1} / (N\bar{x})^N,$$

for $p(\theta) \propto \theta^{-1}$. Further under M_2 the approaches yield

$$f\left(x_j|\hat{\theta}_{i(j)}, M_2\right) = \frac{1}{\bar{x}_{i(j)}} e^{-x_j / \bar{x}_{i(j)}}$$

or

$$\bar{f}\left(x_j|x_{i(j)}^{(N_i-1)}, M_2\right) = \frac{(N_i - 1)\left[(N_i - 1)\bar{x}_{i(j)}\right]^{N_i-1}}{(N_i \bar{x}_i)^{N_i}}$$

where $(N_i - 1)\bar{x}_{i(j)} = N_i\bar{x}_i - x_{ij}$ and $p(\theta_i) \propto \theta_i^{-1}$ for $i = 1, 2$. One then compares

$$\bar{L}_1 = \prod_{j=1}^{N} (N - 1)\left[(N - 1)\bar{x}_{(j)}\right]^{N-1} / (N\bar{x})^N$$

with

$$\bar{L}_2 = \prod_{j=1}^{N_1} \frac{(N_1 - 1)\left[(N_1 - 1)\bar{x}_{1(j)}\right]^{N_1 - 1}}{(N_1\bar{x}_1)^{N_1}}$$

$$\times \prod_{j-N_1+1}^{N} \frac{(N_2 - 1)\left[(N_2 - 1)\bar{x}_{2(j)}\right]^{N_2 - 1}}{(N_2\bar{x}_2)^{N_2}}$$

or compares

$$\hat{L}_1 = \prod_{j=1}^{N} \frac{1}{\bar{x}_{(j)}} e^{-x_j/\bar{x}_{(j)}}$$

with

$$\hat{L}_2 = \prod_{j=1}^{N_1} \frac{1}{\bar{x}_{1(j)}} e^{-x_j/\bar{x}_{1(j)}} \prod_{j=N_1+1}^{N} \frac{1}{\bar{x}_{2(j)}} e^{-x_j/\bar{x}_{2(j)}}.$$

For the Akaike procedure one compares

$$-\log A_1 = N + 1 + N \log \bar{x}$$

with

$$-\log A_2 = N + 2 + N_1 \log \bar{x}_1 + N_2 \log \bar{x}_2.$$

For the Schwarz procedure one compares

$$-\log S_1 = N + \tfrac{1}{2} \log N + N \log \bar{x}$$

with

$$-\log S_2 = N + \log N + N_1 \log \bar{x}_1 + N_2 \log \bar{x}_2.$$

For the third method, using the joint predictive density conditional on x_1, one compares

$$P_1 = \prod_{j=1}^{N-1} \frac{j\left(\sum_{i=1}^{j} x_i\right)^j}{\left(\sum_{i=1}^{j+1} x_i\right)^{j+1}} = \frac{x_1 \Gamma(N)}{(N\bar{x})^N}$$

with

$$P_2 = \prod_{j=1}^{N_1-1} \frac{j\left(\sum_{i=1}^{j} x_i\right)^j}{\left(\sum_{i=1}^{j+1} x_i\right)^{j+1}} \cdot \prod_{j=1}^{N_2-1} \frac{j\left(\sum_{i=1}^{j} x_{N_1+i}\right)^j}{\left(\sum_{i=1}^{j+1} x_{N_1+i}\right)^{j+1}}$$

$$= \frac{x_1 \Gamma(N_1)}{(N_1 \bar{x}_1)^{N_1}} \cdot \frac{x_{N_1+1} \Gamma(N_2)}{(N_2 \bar{x}_2)^{N_2}},$$

the joint density conditional on x_1 and x_{N_1+1}. Although we have used x_1 and x_{N_1+1} here, one could have used in the numerator any x_t and x_k where $1 \le t \le N_1$ and $N_1 + 1 \le k \le N_1 + N_2$. One could average over all choices of x_j, the first observation conditioned on, then

$$\bar{P}_1 = \bar{x} \Gamma(N)/(N\bar{x})^N$$

and similarly averaged over all choices x_t and x_k so that

$$\bar{P}_2 = \bar{x}_1 \bar{x}_2 \Gamma(N_1) \Gamma(N_2) / \left[(N_1 \bar{x}_1)^{N_1} (N_2 \bar{x}_2)^{N_2} \right].$$

Note that in this situation we have violated our rule for starting out at that x_t such that all models have a proper conditional predictive density. We suggest here also the possibility of using

$$\tilde{P}_1 = P_1 f(x_1)$$
$$\tilde{P}_2 = P_2 f(x_1) f(x_{N_1+1})$$

where $f(x)$ is obtained by solving

$$p(\theta)f(x|\theta) = p(\theta|x)f(x)$$

$$\frac{1}{\theta} \cdot \theta e^{-\theta x} = xe^{-\theta x}f(x)$$

or

$$f(x) = \frac{1}{x}.$$

This then results in

$$\tilde{P}_1 = \frac{\Gamma(N)}{(N\bar{x})^N}$$

$$\tilde{P}_2 = \frac{\Gamma(N_1)\Gamma(N_2)}{(N_1\bar{x}_1)^{N_1}(N_2\bar{x}_2)^{N_2}}.$$

P_k, $\bar{P}_k$, and $\tilde{P}_k$ for $i = 1, 2$ are asymptotically equivalent. The use of $\tilde{P}_k$ avoids the necessity of concerning oneself with which x_i is the initial one or of averaging.

Numerical illustration of Example 4.2. The data on the number of correct ledger entries between errors by two bank clerks, excerpted from Geisser and Eddy (1979) and assumed to be approximately exponential, are recorded in Table 4.1.

We now wish to determine whether to consider the two clerks as basically interchangeable with regard to their entry distributions.

Table 4.1   *Correct Ledger Entries between Errors for Clerk 1 and 3*

Clerk 1
734, 121, 404, 646, 1072, 148, 312, 773, 43, 1102, 111, 641, 754
598, 86, 2138, 150, 1047, 907, 165, 166, 6, 94, 1023, 903, 355.

Clerk 3
726, 883, 142, 196, 14, 1905, 456, 2565, 610, 1263, 347, 881, 1214
248, 195, 548, 234, 1096, 530, 338, 356, 217, 195, 77, 392, 3114.

We then calculate the following model selection statistics:

$$\log S_1 = -389.91 \qquad \log A_1 = -388.94 \qquad \log \tilde{P}_1 = -388.99$$

$$\log S_2 = -391.46 \qquad \log A_2 = -389.51 \qquad \log \tilde{P}_2 = -388.92$$

$$\log \bar{L}_1 = -392.84 \qquad \log \hat{L}_1 = -389.02$$

$$\log \bar{L}_2 = -392.41 \qquad \log \hat{L}_2 = -389.60$$

The S_k, A_k, and $\hat{L}_k$ methods favor a single exponential model while the $\tilde{P}_k$ and $\bar{L}_k$ methods favor different exponential models although none of them yields overwhelming evidence in this regard. In this example, then, it is quite likely that with regard to prediction it will not matter very much whether one used the same or separate exponential models.

It is easy to see for this problem that as N_1 and N_2 grow for $\mathcal{L}_k = \bar{L}_k$ or $\hat{L}_k$ that

$$\frac{\mathcal{L}_k}{A_k} \to 1 \quad \text{and for } Q_k = P_k, \tilde{P}_k \text{ or } \tilde{P}_k \text{ that } \frac{Q_k}{S_k} \to 1.$$

This will hold much more generally as shown by Stone (1977) and Dawid (1984).

For the case where we have two normally distributed groups of N_1 and N_2 observations, respectively, whose population means may possibly differ, then a positive value for

$$\log \frac{\tilde{P}_2}{\tilde{P}_1} = \frac{1}{2} \left[\log \frac{N_1 + N_2}{N_1 N_2} + \frac{N_1 N_2 (\bar{x}_1 - \bar{x}_2)^2}{\sigma^2 (N_1 + N_2)} \right]$$

will favor M_2 and a negative value will favor M_1 when the common variance σ^2 is known.

When σ^2 is unknown the calculation of $\log \tilde{P}_2 / \tilde{P}_1$ is as tedious as some of the previous methods, but for sufficiently large N_1 and N_2 the substitution of s^2, the pooled estimate of σ^2, should suffice.

In nested situations for sufficiently large N, the procedures denoted by $\mathcal{L}_k$ and A_k will tend to favor the higher dimensional models as compared with the Q_k and S_k selection methods. The former methods penalize more heavily a lower dimensional error than a higher dimensional error. This can be justifiable if prediction is the ultimate goal.

4.4   Low structure sample reuse model selection

When the formulation of parametric likelihood models is dubious, a more primitive point predictive function can be used to provide guidance on some aspects of modeling. Suppose the possibilities for predicting, using model M_k, imply as previously a predictive function for the next observation

$$x_{N+1,k} = h\left(x^{(N)}, z^{(N)}, z, M_k\right).$$

Then to predict x_j, the jth observed value under M_k using $x_j^{(N-1)}$ we employ

$$\hat{x}_{jk} = h\left(x_j^{(N-1)}, z^{(N)}, M_k\right).$$

The discrepancy for M_k is

$$D_k = \frac{1}{N} \sum_{j=1}^{N} d\left(x_j, \hat{x}_{jk}\right)$$

where $d(\cdot, \cdot)$ represents a defined discrepancy measure between the observed x_j and its predicted value $\hat{x}_{jk}$ under M_k. Low structure "model selection" selects the M_k that optimizes D_k.

Example 4.3. Suppose a set of observations is such wherein each observation has one of two labels and we wish to determine whether the label discriminates with regard to prediction. Let

$$x^{(N_1)} = \left(x_{11}, \ldots, x_{1N_1}\right), \qquad x^{(N_2)} = \left(x_{21}, \ldots, x_{2N_2}\right)$$

and $x^{(N)} = (x^{(N_1)}, x^{(N_2)})$. Then the question arises, given a predictive function, will prediction be better if we treat the set of observations as emanating from one or two populations? Assume the predictive function is $\bar{x}$ the grand mean if the label is irrelevant, i.e., M_1 or $(\bar{x}_1, \bar{x}_2)$, if the label is relevant, i.e., M_2.

Using squared error we have under

$$M_1: \quad D_1 = N^{-1} \sum_{k-1}^{2} \sum_{j=1}^{N_k} \left(x_{kj} - \bar{x}_{(kj)}\right)^2, \qquad (N-1)\bar{x}_{(kj)} = N\bar{x} - x_{kj}$$

and under

$$M_2: \quad D_2 = N^{-1} \sum_{k=1}^{2} \sum_{j=1}^{N_k} \left(x_{kj} - \bar{x}_{k(j)}\right)^2,$$

$$(N_k - 1)\bar{x}_{k(j)} = N_k\bar{x}_k - x_{kj}.$$

Hence we choose M_1 if $D_1 \leq D_2$ and M_2 if $D_1 > D_2$. For $N_1 = N_2 = J \geq 2$, $D_1 > D_2$ is equivalent to choosing M_2 if

$$\frac{J(\bar{x}_1 - \bar{x}_2)^2}{2s^2} > \frac{4J - 3}{2(J-1)}$$

where

$$(N-2)s^2 = \sum_{k=1}^{2} \sum_{j=1}^{J} \left(x_{kj} - \bar{x}_k\right)^2.$$

Applying this method to the illustration of the example of 4.2 involving the two clerks, we obtain $D_1 = 426579$ and $D_2 = 432318$, thus favoring M_1.

If all the x_js were actually i.i.d. and $N(\mu, \sigma^2)$ under M_1 and under

$$M_2, x^{(N_1)} \text{ i.i.d. } N\left(\mu_1, \sigma^2\right) \quad \text{and} \quad x^{(N_2)} \text{ i.i.d. } N\left(\mu_2, \sigma^2\right)$$

then under M_1

$$\frac{J(\bar{x}_1 - \bar{x}_2)^2}{2s^2} \sim F_{1, 2J-2},$$

where $F_{1, 2J-2}$ is an F-variate with 1 and $2J - 2$ degrees of freedom. Hence the probability of choosing M_1 when M_1 is true is

$$\Pr\left[F_{1,2J-2} \le 2 + \frac{1}{2(J-1)}\right] = C_J(R).$$

Note that as J grows $C_J(R) \doteq \Pr[F_{1,2J-2} \le 2]$ and

$$\lim_{J \to \infty} C_J(R) = \Pr[\chi_1^2 \le 2] = 0.843.$$

For this particular situation the Akaike procedure's choice of M_1 is equivalent to

$$f\left(x^{(N)}|\hat{\theta}_2, M_2\right) < ef\left(x^{(N)}|\hat{\theta}_1, M_1\right).$$

Hence it can be calculated that the probability of correctly selecting M_1 is

$$\Pr\left[F_{1,2J-2} < 2(J-1)(e^{1/J} - 1)\right] = C_J(A)$$

$$\lim_{J \to \infty} C_J(A) = \Pr[\chi_1^2 \le 2] = 0.843.$$

For the Schwarz procedure, the probability of correctly choosing M_1 is

$$\Pr\left(F_{1,2J-2} \le 2(J-1)\left[(2J)^{1/2J} - 1\right]\right) = C_J(S),$$

and

$$\lim_{J \to \infty} C_J(S) = 1.$$

Hence only the Schwarz procedure is consistent. For $J \ge 4$,

$$C_J(S) > C_J(A).$$

For $J \ge 6$

$$C_J(S) > C_J(R) > C_J(A).$$

But for B_J representing the probability of correctly selecting M_2

$$B_J(S) < B_J(R) < B_J(A)$$

and

$$\lim_{J \to \infty} B_J(S) = 1.$$

4.5 Predictive model criticism

Here the situation assumed is that one model either conceptually appears to be preeminent in its explanatory or predictive potential or this model has proven adequate for the purposes intended from past experience and initially no other alternatives are entertained until this model's adequacy is sufficiently doubted.

Assume model M is represented by

$$f(x^{(N)}|\theta, M)p(\theta|M) = f(x^{(N)}, \theta|M).$$

Calculate the marginal density of the observations,

$$f(x^{(N)}|M) = \int f(x^{(N)}, \theta|M) \, d\theta.$$

Now Box (1980) suggests that by referring the observed set of values $x^{(N)}$ to the above density we can criticize the model via a significance test. This is basically a predictive significance test, e.g.,

$$\Pr\left[f(X^{(N)}|M) \le f(x^{(N)}|M) \right] = P$$

and when this P-value is sufficiently small, doubts about the adequacy of the model are raised. Sometimes a scalar checking function $T(X^{(N)}) = T$ with observed value $t = t(x^{(N)})$ is used so that

$$\Pr\left[f_T(T|M) \le f_T(t|M) \right] = P.$$

Example 4.4. Let $X^{(N)}$ be a random sample from

$$f(x|\theta) = \theta e^{-\theta x}$$

and

$$p(\theta|\delta,\gamma) = \frac{\gamma^{\delta}\theta^{\delta-1}e^{-\gamma\theta}}{\Gamma(\delta)}.$$

Hence

$$f(x^{(N)}|\delta,\gamma) = \frac{\Gamma(N+\delta)\gamma^{\delta}}{\Gamma(\delta)[x_1 + \cdots + x_N + \gamma]^{N+\delta}}$$

and

$$\Pr[f(X^{(N)}|\delta,\gamma) \leq f(x^{(N)}|\delta,\gamma)] = P$$

$$= \Pr\left[\frac{\Gamma(N+\delta)\gamma^{\delta}}{\Gamma(\delta)[T+\delta]^{N+\delta}} \leq \frac{\Gamma(N+\delta)\gamma^{\delta}}{\Gamma(\delta)(t+\delta)^{N+\delta}}\right] = \Pr[t \leq T]$$

where $T = \Sigma X_i$. Now it is easily seen that

$$f_T(t|\theta) = \frac{\theta^N t^{N-1}e^{-\theta t}}{\Gamma(N)}$$

and

$$f(t|\delta,\gamma) = \int f(t|\theta)p(\theta|\delta,\gamma)\,d\theta$$

$$= \frac{\gamma^{\delta}t^{N-1}}{\Gamma(\delta)\Gamma(N)}\int \theta^{N+\delta-1}e^{-\theta(\gamma+t)}\,d\theta$$

$$= \frac{\gamma^{\delta}\Gamma(N+\delta)t^{N-1}}{\Gamma(N)\Gamma(\delta)(t+\gamma)^{N+\delta}}.$$

Note that

$$U = \frac{\delta}{N}\cdot\frac{T}{\gamma} \sim F_{2N,2\delta}$$

and

$$f_U(u) = \left(\frac{N}{\delta}\right)^N \frac{\Gamma(N+\delta)}{\Gamma(N)\Gamma(\delta)} \frac{u^{N-1}}{[1 + u(N/\delta)]^{N+\delta}}$$

so that if

$$\Pr[U > u] = P,$$

is sufficiently small the model M is in doubt.

Numerical illustration of Example 4.4. We shall apply the results of Example 4.4 to the data on the two clerks in Table 4.1. Suppose past experience indicated that $E(X) \doteq 500$ for clerks. Hence $\gamma = 500(\delta - 1)$, since

$$E(X) = \frac{\gamma}{\delta - 1}, \qquad V(X) = \frac{\delta\gamma^2}{(\delta - 2)(\gamma - 1)^2} \qquad \delta > 2.$$

Suppose now we had no idea of what δ might be and we were not willing to use any particular value of δ. Consequently we can perform the test for a whole range of values of δ and report the resulting levels of significance. Now

$$U_1 = \frac{T_1}{N_1} \frac{\delta}{500(\delta - 1)} = 1.055 \frac{\delta}{\delta - 1} \sim F(26, 2\delta)$$

and

$$U_2 = \frac{T_2}{N_2} \frac{\delta}{500(\delta - 1)} = 1.175 \frac{\delta}{\delta - 1} \sim F(26, 2\delta).$$

For all $\delta > 2$ for Clerk 1, the significance level never falls below 0.4 and for Clerk 3 it never falls below 0.3. In neither case does the test cast doubt on the adequacy of the model.

This approach seems to work well in many cases and requires the existence of $f(x^{(N)}|M)$, which is ensured only when $p(\theta|M)$ is proper. That this procedure is not always sensible is demonstrated in the next example.

Example 4.5. Suppose $X_1, X_2, \ldots,$ is a sequence of i.i.d. Bernoulli variables $P(X_i = 1) = \theta = 1 - P(X_i = 0)$. Suppose we fix the number of trials and let $Y = \sum_1^N X_i$. Now

$$\Pr[Y = y | \theta, M] = \binom{N}{y} \theta^y (1 - \theta)^{N-y}.$$

Let $p(\theta | M) = 1, \theta \in (0, 1)$. Now

$$\Pr[Y = y | M] = \int_0^1 \binom{N}{y} \theta^y (1 - \theta)^{N-y} p(\theta | M)\, d\theta$$

$$= \binom{N}{y} \frac{y!(N-y)!}{(N+1)!} = \frac{1}{N+1}.$$

Hence no model criticism of the kind defined is available.

On the other hand suppose the experiment was terminated as soon as y successes were obtained so that N is random. Further

$$\Pr[N = n | \theta, M] = \binom{n-1}{y-1} \theta^y (1 - \theta)^{n-y} \qquad n = y, y + 1, \ldots$$

and

$$\Pr[N = n | M] = \binom{n-1}{y-1} \int \theta^y (1 - \theta)^{n-y} p(\theta | M)\, d\theta = \frac{y}{n(n+1)}$$

which is a decreasing function of n. Hence model criticism exists for the right tail. Suppose $N = n_0$ is observed then

$$\Pr[N \geq n_0 | M] = \sum_{n=n_0}^{\infty} \frac{y}{n(n+1)} = \frac{y}{n_0} = P.$$

Note that $y/n_0 = \hat{\theta}$, the MLE of θ. Hence the smaller $\hat{\theta}$ the more the model is doubted—but why? Conversely if we stop after a fixed number of failures and $N = n_0$ is observed then it is easy to show

$$\Pr[N \geq n_0 | M] = \sum_{n=n_0}^{\infty} \frac{n_0 - y}{n(n+1)} = 1 - \frac{y}{n_0} = 1 - \hat{\theta} = P.$$

Hence the P-value decreases, as $1 - \hat{\theta}$ decreases or as $\hat{\theta}$ increases. Thus the larger the MLE the more the model is doubted, which is the opposite of the previous case. Now in each of these three cases we have the same probability mechanism, the same prior, the only difference is the stopping rule. So with regard to the model only the stopping rule can be doubted in two of the three cases—but this is absurd. The reason that this happens of course is that the marginal predictive probability function depends strongly on the stopping rule. Hence the likelihood principle has been contradicted and caution should be exercised using the procedure.

4.6 Predictive significance testing for discordancy

Suppose when obtaining an observation x_i, it was noted that some untoward event occurred that may have affected the outcome. An assessment can be made of the discordancy of the observation on the basis of the predictive distribution of X_i (generally prior to observation or independent of the observed value) by a predictive significance test. We calculate, for a suitable region R,

$$\Pr\left[X_i \in R | x_{(i)}, M \right] = P_i,$$

where to ease the notation, $X^{(N)} = (X_i, X_{(i)})$. Now if P_i is small enough one could reject the hypotheses that X_i was concordant with $X_{(i)}$ assuming $X_{(i)}$ concordant with M.

There are two basic ways of implementing the test, i.e., finding the region R of rejection:

1. One could order the values of the range of X_i as to their "scientific" departure from compatability with the rest of the observations, e.g., the further from the center of the predictive distribution the greater the doubt. The center could be $E(X_i|x_{(i)}, M)$ or Mode$(X_i|x_{(i)}, M)$ or Median $(X_i|x_{(i)}, M)$ and R could represent tail areas.
2. One could use the lowest probability density (LPD) region

$$P_i = \Pr\left[f_{X_i}(X_i|x_{(i)}, M) \leq f_{X_i}(x_i|x_{(i)}, M) \right].$$

For a symmetric unimodal predictive probability function both methods will yield the same R in two-sided situations. Even if x_i tests as a discordant or outlying observation one may also be concerned as to the effect it would have on the particular inference or decision in question. Here one can compare the predictive distributions $F_Z(z|x_{(i)})$ and $F_Z(z|x^{(N)})$ of a future observation or set of observations and decide whether or not to include x_i in the inferential analysis. Methods for the comparison will be discussed in a later chapter.

Suppose one did not have any prior suspicion but merely did some data ransacking or dredging with diagnostics of various kinds on the possibility of there being some potentially discordant observations.

We now list various predictive diagnostics that might be used in ransacking the data when dealing with random samples:

3. P_i itself in any of its manifestations can be considered a diagnostic and $\min_i P_i$ is a potential candidate if sufficiently small.
4. The deviation Δ_i of X_i from its predictive mean, mode, or median with the max Δ_i being a candidate (these could be related to P_i through R).
5. The conditional and unconditional predictive ordinates (CPO and UPO) such as

$$d_i = f(x_i|x_{(i)}, M), \qquad u_i = f(x_i|M)$$

and either $\min_i d_i$ or $\min_i u_i$ can be used.
6. Measures of the difference between the predictive distributions of a future observation based on $x^{(N)}$ and $x_{(i)}$ can be used, e.g.,

$$\sup_z \left| F(z|x^{(N)}, M) - F(z|x_{(i)}, M) \right| = S_i.$$

$\text{Max}_i\, S_i$ is then the prime candidate. Another is the Kullback divergence

$$E\left[\log \frac{f(z|x_{(i)}, M)}{f(z|x^{(N)}, M)} \right] = K_i$$

which would yield $\max_i K_i$ as the prime candidate.

Any member of the class of Hellinger distances

$$H_i^n = \int \left| f^{1/n}\left(z|x_{(i)}, M\right) - f^{1/n}\left(z|x^{(N)}, M\right)\right|^n dz$$

can be used and $\max_i H_i^n$ for the particular n is a candidate.

If a review of the generation of the data set did not disclose any currently known or possible circumstance that could have affected the candidate, a statistical determination of its potential discordancy may be in order.

The problem then to be addressed is how to conduct a significance test for such a candidate. Presumably we should take into account how the candidate was chosen by the diagnostic in constructing a test. Say the diagnostic was H and $H(x^{(N)}) = x_C$ was the prime candidate. We then would calculate

$$P_C = \Pr\left[X_C \in R_C | M, H\right]$$

unconditionally where R_C is the region plausibly dictated by the diagnostic's method of choice, i.e., that chose x_C as potentially most discrepant. We shall denote this as an Unconditional Predictive Discordancy (UPD) Test (Geisser 1989, 1990b).

Example 4.6. Let $X_1, \ldots, X_N$ be a random sample from

$$f(x|\theta) = \theta e^{-\theta x}.$$

Suppose the diagnostic led to the plausible choice of

$$x_C = \max_i x_i$$

then

$$\Pr(X_C > x_C | M, \theta, C) = 1 - \left(1 - e^{-x_C \theta}\right)^N$$

$$P_C = \Pr(X_C > x_C | M, C) = 1 - \int \left(1 - e^{-x_C \theta}\right)^N p(\theta)\, d\theta.$$

Now suppose

$$p(\theta) \propto \theta^{N_0 - 1} e^{-\theta N_0 \bar{x}_0}$$

then

$$P_C = \sum_{j=1}^{N} \binom{N}{j} (-1)^{j+1} \left(\frac{N_0 \bar{x}_0}{N_0 \bar{x}_0 + j x_C} \right)^{N_0}.$$

Example 4.7. Let $X_1, \ldots, X_N$ be a random sample from $N(\theta, 1)$. Assume θ is $N(\beta, \tau^2)$ where β and τ^2 are presumed known. Now the marginal distribution of X_i is $N(\beta, 1 + \tau^2)$. Further the predictive distribution of Z is $N(a, b_0^2)$ where

$$a = \frac{\tau^2 \bar{x} + (1/N)^{\beta}}{\tau^2 + N^{-1}}, \qquad b_0^2 = 1 + \tau^2 (N\tau^2 + 1)^{-1}.$$

Hence for Z based on $x_{(i)}$ rather than $x^{(N)}$, Z is $N(a_i, b^2)$,

$$a_i = \frac{\tau^2 \bar{x}_{(i)} + [1/(N-1)]\beta}{\tau^2 + [1/(N-1)]}, \qquad b^2 = 1 + \tau^2 [1 + (N-1)\tau^2]^{-1},$$

and $(N-1)\bar{x}_{(i)} = \sum_{j \neq i} x_j$.

In this instance all the unconditional discordancy indices previously discussed select the x_i that maximizes $|x_i - \beta|$ or the checking function

$$v_i^2 = \frac{(x_i - \beta)^2}{1 + \tau^2}.$$

If a particular x_i had been flagged prior to observation, then

$$P_i = \Pr\left(\chi_1^2 > \frac{(x_i - \beta)^2}{1 + \tau^2} \right).$$

To implement the unconditional test based on ransacking the data we note that $X = (X_1, \ldots, X_N)'$ is distributed unconditionally as a multivariate $N(\mu, \Sigma)$ where $\mu = \beta e$, $e = (1, \ldots, 1)'$, and Σ is a known matrix with equal diagonal elements $1 + \tau^2$ and equal off diagonal elements. One then would find the distribution of the $\max |X_i - \beta|$ or $\max(X_i - \beta)^2/(1 + \tau^2) = V_C^2$, i.e., the maximum of

N correlated χ_1^2 variates. The calculation of

$$\Pr\left[V_C^2 \geq v_C^2\right] = P_C$$

would then result in the significance level for the UPD test. It is clear that even for such a simple normal case, the calculation can be quite arduous, especially in cases where the variance is also unknown.

Clearly a UPD test will not exist when improper priors are used. Since these priors are often quite useful in Bayesian analysis, we propose a procedure for dealing with this situation.

A way of approaching the problem is to consider the predictive distribution of the choice of the diagnostic $Z(x_j)$, where say the diagnostic orders the observations from least discrepant to most discrepant, $j = 1, \ldots, C$. We then calculate the ordinary predictive distribution of X_C and $Z(X_C)$ conditional on the rest of the observations and adjust by the following calculation

$$P_C = \Pr\left[Z \geq x_C | Z \geq x_{C-1}, M, x_{(C)}\right],$$

where x_{C-1} represents the diagnostic's choice of the second most discrepant observation and where larger values imply greater discrepancy. Then one determines whether it is sufficiently removed from the next most discrepant value. We denote this as an adjusted Conditional Predictive Discordancy Test or simply a CPD test.

Example 4.6 (*continued*). Let $X^{(N)}$ be a random sample from the exponential distribution with prior density as in Example 4.6, then

$$P_i = \Pr\left[X_i \geq x_i | x_{(i)}, M\right] = \left[\frac{N_0 \bar{x}_0 + N\bar{x}}{N_0 \bar{x}_0 + N\bar{x} + x_i}\right]^{N+N_0}.$$

Of course this would be inappropriate after ransacking. One way of adjusting this to give a sensible answer is to calculate

$$P_C = \Pr\left[Z > x_c | Z > x_{C-1}, x_{(C)}\right],$$

where x_C and x_{C-1} are largest and second largest observations and $x_{(C)}$ denotes the set of observations with x_C deleted, since all sensible discordancy indices would lead to the largest as the potentially discordant candidate.

The significance level for the proposed adjusted or Conditional Predictive Discordancy (CPD) test for the exponential problem previously discussed is

$$P_C = \left[\frac{N_0 \bar{x}_0 + N\bar{x} + x_{C-1} - x_C}{N_0 \bar{x}_0 + N\bar{x}} \right]^{N_0 + N - 1},$$

which depends not only on the prior hyperparameters N_0 and $\bar{x}_0$ but on the observables x_C, x_{C-1}, $\bar{x}$, and N. We also note that the test exists as $N_0 \to 0$, which yields the useful improper prior that purports to reflect little prior information relative to the likelihood. For example, $N_0 \to 0$ results in

$$p(\theta) \propto \theta^{-1}$$

and

$$\lim_{N_0 \to 0} P_C = \left[\frac{N\bar{x} + x_{C-1} - x_C}{N\bar{x}} \right]^{N-1}. \tag{4.1}$$

This, it is easy to show, leads to the same significance level one would obtain from the frequentist statistic

$$W = \frac{X_C - X_{C-1}}{N\bar{X}}$$

that would be used to test for discordance, so that the frequentist calculation

$$\Pr[W > w] = P_C.$$

is equivalent to Eq. (4.1).

However the usual frequentist approach is to calculate the frequentist significance level for

$$T = (X_C - X_{C-1})/X_C,$$

which is

$$\Pr[T > t] = N(N-1)B\left(\frac{2-t}{1-t}, N-1\right)$$

when $B(\cdot, \cdot)$ is the beta function (Dixon 1950, 1951; Likes, 1966).

Example 4. 7 (*continued*). For the normal case of Example 4.7 we suggest using the CPD approach to test discordancy and for the subsequent normal sampling that we shall discuss. For Example 4.7, conditional on $x_{(i)}$, the checking function was

$$V^2 = \left(\frac{Z - a_i}{b}\right)^2 \sim \chi_1^2.$$

A conditional significance test can be calculated as

$$\Pr\left[V^2 \ge v_C^2 | V^2 \ge v_{C-1}^2, x_{(C)}\right] = P_C$$

or

$$P_C = \frac{1 - F(v_C^2)}{1 - F(v_{C-1}^2)}$$

where $F(\cdot)$ is the distribution function of a χ_1^2 random variable. The result then is immediately available from the χ_1^2 distribution.

Example 4.8. A particular case of greater interest is when the random sample is from a $N(\mu, \sigma^2)$ and μ and σ^2 are unknown. To make this situation closely correspond to a conventional frequency analysis we let the prior density be

$$p(\mu, \sigma^2) \propto 1/\sigma^2,$$

noting that now the UPD test is unavailable. Again computation of any of the reasonable discordancy indices will lead to the selection of that x_i that maximizes $(x_i - \bar{x}_{(i)})^2$ or the checking function

$$\frac{(x_i - \bar{x}_{(i)})^2 (N-1)}{N s_{(i)}^2}$$

which will yield the same x_C, since $s_{(C)}^2 \le s_{(i)}^2$ where $(N-1)\bar{x}_{(i)} = \sum_{j \ne i} x_j$ and $(N-2)s_{(i)}^2 = \sum_{j \ne i}(x_j - \bar{x}_{(i)})^2$.

We note that the predictive distribution of the checking function

$$W^2 = \frac{\left(Z - \bar{x}_{(i)}\right)^2 (N-1)}{Ns_{(i)}^2}$$

conditional on $x_{(i)}$ has an $F_{1, N-2}$ distribution. Hence to test for discordancy of x_C we can calculate the conditional or adjusted significance level

$$P_C = \Pr\left[W^2 > w_C^2 | W^2 > w_{C-1}^2, x_{(C)} \right]$$

$$= \frac{1 - F(w_C^2)}{1 - F(w_{C-1}^2)}$$

where w_{C-1}^2 is the observed second largest standardized deviate and $F(\cdot)$ is the distribution of an F variate with 1 and $N-2$ degrees of freedom.

Numerical illustration of Example 4.8. As an application of Example 4.8, consider the following experiment (Barnett and Lewis, 1978) on crushing strength of cement made up of 10 test cubes from a particular mix. After a suitable hardening period the observed strengths of the cubes in pressure per square inch is determined to be 790, 750, 910, 650, 990, 630, 1290, 820, 860, 710. We assume, as the aforementioned authors, that previous experience has indicated that the normal distribution is adequate for crushing strength observables. The maximum discordancy index occurs obviously for the value 1290 resulting in $w_C^2 = 15.873$, with the second largest for the value 630, yielding $w_{C-1}^2 = 1.827$. Hence $P_C = 0.02$.

Example 4.6 (*continued*). In this elaboration of the previous example we now allow for censoring but restrict ourselves to the noninformative prior. We are dealing with independent copies of exponentially distributed random variables. Now suppose that

$X_1, \ldots, X_d$ are fully observed and $X_{d+1}, \ldots, X_N$ are censored at $x_{d+1}, \ldots, x_N$. Let x_C and x_{C-1} be the largest and second largest observations, respectively. Calculate the usual predictive distribution of Z based on $x_{(C)}$ and then compute

$$\Pr\left[Z \geq x_C \mid Z > x_{C-1}, x_{(C)} \right] = \left[\frac{x_{C-1} + (N-1)\bar{x}_{(C)}}{x_C + (N-1)\bar{x}_{(C)}} \right]^c$$

where $(N-1)\bar{x}_{(C)}$ is the sum of the observations with x_C deleted and for $d \leq N - 1$, $c = d$ or $d - 1$ depending on whether x_C is censored or not and $c = N - 1$ if $d = N$. For further elaborations on this problem see Geisser (1989).

Numerical illustration of Example 4.6. We use the data on the number of correct ledger entries between errors by bank clerks given in the illustration of Example 4.2. The largest entry, 2138, is tested as a possibly discordant observation by

$$P_C = \left(\frac{1102 + 12{,}361}{2138 + 12{,}361} \right)^{25} = 0.16.$$

The P_C value does not cause us concern as to the possible discordancy of the entry 2138. If suspicion about the discordancy of this value was for reasons other than it was the largest, then we could be concerned because

$$P_i = \left(\frac{14{,}499}{14{,}499 + 2138} \right)^{25} = 0.03.$$

Example 4.9. Using the linear regression setup of Chapter 3, Section 11, the predictive density of

$$\frac{\left(Y_i - x_i' \hat{\beta}_{(i)} \right)^2}{(N - 1 - p)s_{(i)}^2 \left[1 + x_i'(X_{(i)}' X_{(i)})^{-1} x_i \right]} = U^2$$

is distributed as an $F_{1, N-1-p}$ variate where

$$x_i' = (x_{i1}, \ldots, x_{ip}), \qquad y' = (y_i, y_{(i)}')$$

$$X_{(i)}' = (x_1, \ldots, x_{i-1}, x_{i+1}, \ldots, x_N)$$

$$\hat{\beta}_{(i)} = (X_{(i)}' X_{(i)})^{-1} X_{(i)}' y_{(i)}$$

$$(N - 1 - p) s_{(i)}^2 = (y_{(i)} - X_{(i)} \hat{\beta}_{(i)})'(y_{(i)} - X_{(i)} \hat{\beta}_{(i)}).$$

Hence a CPD test for a previously suspicious Y_i and associated x_i would entail calculating

$$P = \Pr[U^2 > u^2 | y_{(i)}] = 1 - F(u^2)$$

where $F(\cdot)$ is the $F_{1, N-1-p}$ distribution function. If on the other hand Y_C is flagged by a diagnostic then for the adjusted CPD test we calculate

$$P_C = \Pr[U^2 > u_C^2 | U^2 > u_{C-1}^2, u_{(C)}^2] = \frac{1 - F(u_C^2)}{1 - F(u_{C-1}^2)}$$

where u_C^2 and u_{C-1}^2 are largest and second largest values of u_i^2, $i = 1, \ldots, N$. For a numerical illustration see Geisser (1987b).

Discordancy testing procedures can be developed from both the conditional and unconditional approach (Geisser, 1989, 1990b). There are several aspects of the conditional approach that many statisticians may find appealing. The first is that the test procedures depend more on the conditional distribution (likelihood) of the observables than the unconditional procedures, which depend more heavily on the prior for θ. The second is that the conditional approach can be used either with proper or certain useful improper priors, while the unconditional requires a proper prior. A third aspect is that the conditional approach will often lead to much simpler computations than the unconditional approach.

It is to be noted that all these tests are essentially subjective assessments and are not grounded in a frequency theory as there is no attempt to tie this to a class of repetitions.

References

Akaike, H. (1973). Information theory and an extension of the maximum likelihood principle. In *Proceedings of the 2nd International Symposium on Information Theory*, B. N. Petrov and F. Czaki (eds.). Budapest: Akademiai Kiado, 267–281.

Atkinson, A. C. (1980). A note on the generalized information criterion for choice of a model. *Biometrika* 67(2), 413–418.

Barnett, V., and Lewis, T. (1978). *Outliers in Statistical Data*. New York: Wiley.

Box, G. E. P. (1980). Sampling and Bayes' inference in scientific modelling and robustness. *Journal of the Royal Statistical Society B* 143, 383–430.

Dawid, A. P. (1984). Statistical theory, the prequential approach. *Journal of the Royal Statistical Society A* 147(2), 278–292.

Dixon, W. J. (1950). Analysis of extreme values. *Annals of Mathematical Statistics* 21, 488–506.

Dixon, W. J. (1951). Ratios involving extreme values. *Annals of Mathematical Statistics* 22, 68–78.

Geisser, S. (1987b). Influential observations, diagnostics and discordancy tests. *Journal of Applied Statistics* 14(2), 133–142.

Geisser, S. (1989). Predictive discordancy tests for exponential observations. *The Canadian Journal of Statistics* 17(1), 19–26.

Geisser, S. (1990b). Predictive approaches to discordancy testing. In *Bayesian and Likelihood Methods in Statistics and Econometrics*, S. Geisser et al. (eds.). Amsterdam: North-Holland, 321–335.

Geisser, S., and W. F. Eddy (1979). A predictive approach to model selection. *Journal of the American Statistical Association*, 14, 153–160.

Leonard, T. (1982). Comment. *Journal of the American Statistical Association* 77(379), 657–658.

Likes, J. (1966). Distribution of Dixon's statistics in the case of an exponential population. *Metrika* 11, 46–54.

San Martini, A., and Spezzaferri, F. (1984). A predictive model selection criterion. *Journal of the Royal Statistical Society B* 46(2), 296–303.

Schwarz, G. (1978). Estimating the dimension of a model. *Annals of Statistics* 6, 461–464.

Smith, A. F. M., and Spiegelhalter, D. J. (1980). Bayes factors and choice criteria for linear models. *Journal of the Royal Statistical Society B* 42, 213–220.

Stone, M. (1977). An asymptotic equivalence of choice of model by cross-validation and Akaike's criterion. *Journal of the Royal Statistical Society B* 39, 44–47.

CHAPTER 5

Problems of comparison and allocation

A large number of problems in statistics involves comparisons of different populations or the same population subjected to different treatments. In classical statistics, these problems are framed in the rubric of hypothesis testing and estimation of parameters. The conventional Bayesian approach is similar but differs in calculating either posterior odds of various parametric hypotheses or posterior distribution of parametric functions. The following is typical: Let two independent random samples $X^{(N_1)}$ and $Y^{(N_2)}$ be drawn from populations having possible different means μ and η say. A standard practice is to make a comparison of μ and η or in particular a posterior probability region interval on $\mu - \eta$.

5.1 Comparisons

We shall focus on predictive comparisons of future sets $X_{(M)}$ and $Y_{(M)}$. For example, particular interest may be focused on a single future comparison,

$$\Pr\left[X_{N_1+1} - Y_{N_2+1} \le u \right];$$

or when dealing with positive random variables

$$\Pr\left[X_{N_1+1} / Y_{N_2+1} \le u \right];$$

or more generally on some probability comparison of $h(X_{(M)})$ with $h(Y_{(M)})$ (Geisser, 1971).

Example 5.1. Consider that the $X^{(N_1)}$ are i.i.d. $N(\mu, 1)$ and the $Y^{(N_2)}$ are i.i.d. $N(\eta, 1)$ and let $p(\mu, \eta) = p(\mu)p(\eta) = \text{const.}$
Hence

$$X_{N_1+1}|x^{(N_1)} \sim N\left(\bar{x}_{N_1}, 1 + \frac{1}{N_1}\right),$$

$$Y_{N_2+1}|y^{(N_2)} \sim N\left(\bar{y}_{N_2}, 1 + \frac{1}{N_2}\right)$$

and

$$Z = X_{N_1+1} - Y_{N_2+1} \sim N\left(\bar{x}_{N_1} - \bar{y}_{N_2}, 2 + \frac{1}{N_1} + \frac{1}{N_2}\right).$$

Then

$$\Pr\left[X_{N_1+1} - Y_{N_2+1} \le u\right] = \Phi\left(\frac{u - \left(\bar{x}_{N_1} - \bar{y}_{N_2}\right)}{\sqrt{2 + \dfrac{1}{N_1} + \dfrac{1}{N_2}}}\right)$$

and, e.g., $u = 0$ yields $\Pr[X_{N_1+1} \le Y_{N_2+1}]$ as a single comparative measure but the whole distribution of Z is the complete informative comparison.

Further if we define

$$M\bar{\bar{X}}_M = \sum_{i=1}^{M} X_{N_1+i}, \quad M\bar{\bar{Y}}_M = \sum_{i=1}^{M} Y_{N_2+i}$$

then for $e' = (1, \ldots, 1)$,

$$\begin{pmatrix} X_{N_1+1} \\ \vdots \\ X_{N_1+M} \end{pmatrix} \sim N\left(\bar{x}_{N_1}e, I + N_1^{-1}ee'\right),$$

so that

$$E\left(\bar{\bar{X}}_M\right) = \bar{x}_{N_1}$$

and

$$\text{Var}\left(\bar{\bar{x}}_M\right) = \frac{M(N_1+1) + M(M-1)}{N_1 M^2} = \frac{N_1+M}{N_1 M}.$$

Hence, after a similar calculation for $\bar{Y}_M$,

$$\bar{\bar{X}}_M - \bar{\bar{Y}}_M \sim N\left(\bar{x}_{N_1} - \bar{y}_{N_2}, \frac{1}{N_1} + \frac{1}{N_2} + \frac{2}{M}\right).$$

Note now as M grows

$$\bar{\bar{X}}_M - \bar{\bar{Y}}_M \to \mu - \eta \sim N\left(\bar{x}_{N_1} - \bar{y}_{N_2}, \frac{1}{N_1} + \frac{1}{N_2}\right)$$

so that $\mu - \eta$ can be justified as a limit of a particular interesting function of observables and could provide a normative evaluation or an approximation for large M for the comparison of future averages.

Example 5.2. In the previous example let the Xs and Ys have unknown but common variance. Suppose we assume in addition that

$$p(\sigma^2) \propto \frac{1}{\sigma^2}.$$

Then for arbitrary M, it is easy to show that the predictive distribution of

$$\frac{\left(\bar{\bar{X}}_M - \bar{\bar{Y}}_M\right) - \left(\bar{x}_{N_1} - \bar{y}_{N_2}\right)}{[(1/N_1) + (1/N_2) + (2/M)]s^2}$$

is Student's t with $N_1 + N_2 - 2$ degrees of freedom, where

$$(N_1 + N_2 - 2)s^2 = \sum_i^{N_1} (x_i - \bar{x}_{N_1})^2 + \sum_i^{N_2} (y_i - \bar{y}_{N_2})^2.$$

Now the comparison depends on the t distribution. Note $M = 1$ and $M \to \infty$ are the two most interesting cases, with the scaling factor for s^2 going from $N_1^{-1} + N_2^{-1} + 2M^{-1}$ to a minimum of $N_1^{-1} + N_2^{-1}$.

Example 5.3. More generally, if we have K groups then one should be interested in the ordering of the groups. Let X_{ij} be independent and

$$N(\mu_i, \sigma^2) \qquad i = 1, \ldots, K \qquad j = 1, \ldots, N_i.$$

Then we need to compute the predictive probabilities of all permutations,

$$P\left(X_{i_1, N_{i_1}+1} \leq X_{i_2, N_{i_2}+1} \leq \cdots \leq X_{i_K, N_{i_K}+1} \right),$$

when $i_1, \ldots, i_K$ is a permutation of the first K positive integers. If $K = 3$, there are six comparative probabilities.

If we let $X_{i, N_i+1} = Z_i$, then we may calculate a more informative analysis than the usual ANOVA. Here

$$\Pr[Z_1 \leq Z_2 \leq Z_3] = P_{123}$$
$$\Pr[Z_1 \leq Z_3 \leq Z_2] = P_{132}$$
$$\Pr[Z_2 \leq Z_1 \leq Z_3] = P_{213}$$
$$\Pr[Z_2 \leq Z_3 \leq Z_1] = P_{231}$$
$$\Pr[Z_3 \leq Z_1 \leq Z_2] = P_{312}$$
$$\Pr[Z_3 \leq Z_2 \leq Z_1] = P_{321}$$

If these are all rather close, i.e., close to $1/6$, then we may conclude that it is not worth making a distinction between them. However, if there are differences we would choose, all other things being equal,

$$\max_{i_1,i_2,i_3} P_{i_1,i_2,i_3}$$

as the best ordering. If we were interested in the best group, then we compare

$$P_{123} + P_{213} = \text{Probability that } Z_3 = \text{Max}(Z_1, Z_2, Z_3)$$
$$P_{132} + P_{312} = \text{Probability that } Z_2 = \text{Max}(Z_1, Z_2, Z_3)$$
$$P_{231} + P_{321} = \text{Probability that } Z_1 = \text{Max}(Z_1, Z_2, Z_3).$$

If $p(\mu_1, \mu_2, \mu_3, \sigma^2) \propto 1/\sigma^2$, then one can easily obtain the joint predictive density of $Z' = (Z_1, Z_2, Z_3)$,

$$f(z_1, z_2, z_3) \propto \frac{1}{\left[a + \sum_i \dfrac{N_i}{N_i + 1} (z_i - \bar{x}_i)^2 \right]^{N/2}}$$

where $a = \sum_i (N_i - 1)s_i^2$, $N\bar{x}_i = \sum_{j=1}^{N_i} x_{ij}$, and $N = \sum_i N_i$. Setting $z' = (z_1, z_2, z_3)$ and $\bar{x}' = (\bar{x}_1, \bar{x}_2, \bar{x}_3)$ then $Z \sim S_3(N - 3, \bar{x}, \Sigma)$ where $\Sigma = \{\sigma_{ij}\}$ and $\sigma_{ij} = a(N_i + 1)/N_i$ for $i = j$, and 0 otherwise. The various probabilities can be numerically calculated from the above student distribution.

Numerical illustration of Example 5.3. Three different brands of gasoline with the same octane rating were tested on a 1000 mile run of a particular make of utility vehicle. Brand i was tested on N_i vehicles. The data are given in average miles per gallon for each brand in Table 5.1.

We assume normality of the observable miles per gallon variable, equality of variances for all brands, and vague prior knowledge about μ and σ^2. Letting Z_i denote the miles per gallon

Table 5.1   *Sample Means and Variances for Brands*

Brand	N_i	$\bar{x}_i$	s_i^2	a
1	8	12.5395	0.2667	21.0852
2	10	13.3277	0.9959	
3	13	14.4082	0.8546	

variable for brand i, $i = 1, 2, 3$, we obtain the predictive probabilities of the ordering of the brands in miles per gallon: $P_{123} = 0.2060$, $P_{213} = 0.1872$, $P_{132} = 0.1868$, $P_{231} = 0.1420$, $P_{312} = 0.1460$, and $P_{321} = 0.1320$.

It is of some interest to point out that if we use the posterior distribution of $\mu = (\mu_1, \mu_2, \mu_3)$ that is $S_3(N - 3, \bar{x}, \Sigma)$ where $\sigma_{ij} = N_i^{-1}a$, for $i = j$ and 0 otherwise for ordering the population brand means, we obtain $P'_{123} = 0.3404$, $P'_{213} = 0.2304$, $P'_{132} = 0.2232$, $P'_{231} = 0.1236$, $P'_{312} = 0.0356$, $P'_{321} = 0.0468$. As expected, the brands are still in the same probability order. The latter set of probabilities is much more spread out than the former. The latter concern is with the average for a very large fleet of vehicles. The former is for a single vehicle. Further let $V = \max_i Z_i$, then

$$\Pr[V = Z_3] = 0.393, \qquad \Pr[V = Z_2] = 0.333, \qquad \text{and}$$
$$\Pr[V = Z_1] = 0.274$$

while for $\gamma = \max(\mu_i)$ we obtain

$$\Pr(\gamma = \mu_3) = 0.571, \qquad \Pr(\gamma = \mu_2) = 0.259, \qquad \text{and}$$
$$\Pr(\gamma = \mu_1) = 0.170$$

which illustrates the sharper results obtained for a mean as opposed to a single future observable.

Example 5.4. If we are dealing with three exponential populations, say, then the computation can be easily made. Let i.i.d.

$$X_{ij}, i = 1, 2, 3, \qquad j = 1, \ldots, N_i, \qquad \text{and} \qquad f(x_i | \theta_i) = \theta_i e^{-\theta_i x_i}.$$

Let $Z_i = X_{i, N_i + 1}$, then to calculate the probability that, say, Z_3 is the largest we first set $V = \text{Max}(Z_1, Z_2)$. Then

$$\Pr[V \le v | \theta] = (1 - e^{-\theta_1 v})(1 - e^{-\theta_2 v})$$
$$= 1 - e^{-\theta_1 v} - e^{-\theta_2 v} + e^{-v(\theta_1 + \theta_2)}$$
$$f(v | \theta) = \theta_1 e^{-\theta_1 v} + \theta_2 e^{-\theta_2 v} - (\theta_1 + \theta_2)e^{-v(\theta_1 + \theta_2)}$$
$$\Pr[Z_3 \ge V | \theta] = \int_0^\infty \left[\theta_1 e^{-\theta_1 v} + \theta_2 e^{-\theta_2 v} - (\theta_1 + \theta_2)e^{-v(\theta_1 + \theta_2)} \right]$$
$$\times e^{-\theta_3 v} dv$$
$$= \frac{\theta_1 \theta_2 (\theta_1 + \theta_2 + 2\theta_3)}{(\theta_1 + \theta_3)(\theta_2 + \theta_3)(\theta_1 + \theta_2 + \theta_3)}.$$

Hence

$$\Pr[Z_3 \ge V] = \iiint \frac{\theta_1 \theta_2 (\theta_1 + \theta_2 + 2\theta_3)}{(\theta_1 + \theta_3)(\theta_2 + \theta_3)(\theta_1 + \theta_2 + \theta_3)}$$
$$\times p(\theta_1, \theta_2, \theta_3 | D) \, d\theta$$

where

$$D = (x^{(N_1)}, x^{(N_2)}, x^{(N_3)})$$

and

$$p(\theta_1, \theta_2, \theta_3 | D) \propto L(D | \theta) p(\theta)$$

where $\theta = (\theta_1, \theta_2, \theta_3)$. Using

$$p(\theta) \propto \prod_{i=1}^{3} \theta_i^{-1},$$

$$p(\theta_1, \theta_2, \theta_3 | D) \propto \prod_{i=1}^{3} \theta_i^{N_i - 1} e^{-\theta_i (N_i \bar{x}_i)}$$

then the triple integral can be calculated numerically. If we substitute $1/\bar{x}_i = \hat{\theta}_i$ we can get an initial estimate of

$$\Pr[Z_3 \ge V] \doteq \frac{\bar{x}_3^2 (\bar{x}_2 \bar{x}_3 + \bar{x}_1 \bar{x}_3 + 2\bar{x}_1 \bar{x}_2)}{(\bar{x}_1 + \bar{x}_3)(\bar{x}_2 + \bar{x}_3)(\bar{x}_1 \bar{x}_2 + \bar{x}_1 \bar{x}_3 + \bar{x}_2 \bar{x}_3)}.$$

Numerical illustration of Example 5.4. Suppose in addition to Clerk 1 and Clerk 3 of Table 4.1 we also have data on Clerk 5 (Geisser and Eddy, 1979) (Table 5.2). Using the notation Z_i for a future value for Clerk i, $i = 1, 3, 5$ we calculate by numerical integration the probability that a future value for Clerk 3 will be

Table 5.2   *Correct Ledger Entries between Errors for Clerk 5*

149, 74, 170, 2, 129, 3, 65, 44, 204, 333, 60, 11, 60
20, 608, 19, 64, 113, 413, 75, 22, 403, 299, 396, 6, 156.

the largest to be 0.5297, which is the same to two decimal places as the initial estimate 0.5325.

Example 5.5. Another type of a two group comparison when the observables must be of the same sign may be formulated in terms of

$$\Pr[Z_1 > tZ_2]$$

$$= \iint p(\theta_1, \theta_2 | D) \int \Pr[Z_1 > tz_2 | \theta_1, \theta_2] f(z_2 | \theta_1, \theta_2) \, dz_2 \, d\theta_1 \, d\theta_2$$

For the previously discussed exponential case

$$\Pr[Z_1 > tZ_2] = \int p(\theta_1, \theta_2 | D) \int e^{-t\theta_1 z_2} \theta_2 e^{-\theta_2 z_2} \, dz_2 \, d\theta_1 \, d\theta_2$$

$$= \iint p(\theta_1, \theta_2 | D) \frac{\theta_2}{t\theta_1 + \theta_2} \, d\theta_1 \, d\theta_2.$$

This may also be evaluated numerically for any given t. However, it can be shown that

$$\frac{(N_2 - 1)\bar{x}_1}{(N_2 - 1)\bar{x}_1 + tN_2\bar{x}_2} \le P(Z_1 > tZ_2) \le \frac{N_1\bar{x}_1}{N_1\bar{x}_1 + t(N_1 - 1)\bar{x}_2}$$

using Jensen's inequality and the fact that $N_1\bar{x}_1\theta_1(N_2\bar{x}_2\theta_2)^{-1}$ can be transformed to a beta variate. Hence

$$P(Z_1 > tZ_2) \doteq \frac{\bar{x}_1}{\bar{x}_1 + t\bar{x}_2}$$

since the quantity above on the right is between the bounds for all N_1 and N_2 and is the value that both the lower and upper bounds approach.

Numerical illustration of Example 5.5. We calculate the probability that a future value of Clerk 3 will exceed that of Clerk 5 by factors say $t = 1$ and $t = 2$, and that value $t = \bar{t}$ such that

$$\Pr(Z_1 > \bar{t}Z_2) = \frac{1}{2}.$$

By numerical integration we obtain probabilities 0.824 and 0.703 for $t = 1$ and $t = 2$, respectively. Note that the approximations yield 0.828 and 0.706. Further numerical integration yields $\bar{t} = 4.8$.

5.2 Allocation

Sometimes the problem is to allocate a new observation to a treatment to obtain some optimal result where randomly selected groups from a population are each assigned a single treatment. Here the result is required to be as close as possible to some given value say a or to be in a particular interval or region. The first criterion could indicate allocating treatment j to a future individual Z, such that treatment j is the solution to

$$\operatorname*{Min}_{i} \int (z - a)^2 f(z|i, x)\, dz$$

where $f(z|i, x)$ is the predictive density of a future observation given data x, that is to be allocated treatment i. Another often useful criterion is to choose that j that is the solution to

$$\operatorname*{Max}_{i} P(Z \in R|i, x)$$

where R is the region of interest.

Example 5.1 *continued.* Suppose what is required here is to have the future value as close as possible in terms of squared error to zero. Hence we would allocate a new observation Z to that treatment X or Y that yielded the smaller of

$$E\left(X^2 | \bar{x}_{N_1}, 1 + \frac{1}{N_1} \right) = 1 + \frac{1}{N_1} + \bar{x}_{N_1}^2$$

$$E\left(Y^2 | \bar{y}_{N_1}, 1 + \frac{1}{N_2} \right) = 1 + \frac{1}{N_2} + \bar{y}_{N_2}^2.$$

If we required the largest probability that $Z \in (-b, b)$ for $b > 0$ then we would allocate Z to treatment X or Y for whichever of the following

$$\Phi\left(\frac{b - \bar{x}_{N_1}}{\sqrt{1 + N_1^{-1}}}\right) - \Phi\left(\frac{-b - \bar{x}_{N_1}}{\sqrt{1 + N_1^{-1}}}\right),$$

$$\Phi\left(\frac{b - \bar{y}_{N_2}}{\sqrt{1 + N_2^{-1}}}\right) - \Phi\left(\frac{-b - \bar{y}_{N_2}}{\sqrt{1 + N_2^{-1}}}\right)$$

was the larger.

Example 5.6. Suppose we were required to allocate the next observation Z to treatment 1 or treatment 2 so that the result will be as close as possible to the value a. Assume

$$X_{11}, \ldots, X_{1N_1} \text{ are } i.i.d. \ N(\mu_1, \sigma_1^2)$$

$$Y_{21}, \ldots, X_{2N_2} \text{ are } i.i.d. \ N(\mu_2, \sigma_2^2)$$

$$\prod_{i=1}^{2} p(\mu_i, \sigma_i^2) \propto \frac{1}{\sigma_1^2 \sigma_2^2}.$$

Then for $N_i \bar{x}_i = \sum_{j=1}^{N_i} x_{ij}$ and $(N_i - 1)s_i^2 = \sum_{j=1}^{N_i}(x_{ij} - \bar{x}_i)^2$

$$Z_i \sim S_1\left(N_i - 1, \bar{x}_i, s_i^2(N_i^2 - 1)/N_i\right) \qquad i = 1, 2$$

and we would allocate that treatment i to Z that minimizes

$$E(Z_i - a)^2 = \text{Var}(Z_i) + (\bar{x}_i - a)^2$$

$$= s_i^2\left(\frac{N_i^2 - 1}{N_i(N_i - 3)}\right) + (\bar{x}_i - a)^2 \qquad N_i > 3.$$

On the other hand if we required the largest probability such that $Z \in (a - b, a + b)$ we would allocate that treatment i to Z for whichever i maximized

$$p_i = S\left(\frac{(a + b - \bar{x}_i)\sqrt{N_i}}{s_i\sqrt{1 + N_i}}\right) - S\left(\frac{(a - b - x_i + \bar{x}_i)\sqrt{N_i}}{s_i\sqrt{1 + N_i}}\right)$$

where $S(\cdot)$ presents the student distribution function with $N_i - 1$ degrees of freedom.

Numerical illustration of Example 5.6. An expensive machine part is to be purchased from one of two manufacturers, A and B. Both manufacturers charge the same amount for the part. The standard length for a part is 1000 mm and it will work efficiently between 998 and 1002 mm. A sample of size 10 with mean length 997.38 mm and variance 1.462 mm^2 is available from A and a sample of 15 with mean length 1001.50 mm and variance 8.760 mm^2 is available from B.

If one wants the part to be as close as possible to 1000 mm then

$$E(Z_1 - 1000)^2 = \frac{81}{70} \times 1.462 + (997.38 - 1000)^2 = 8.556$$

$$E(Z_2 - 1000)^2 = \frac{324}{180} \times 8.760 + (1001.5 - 1000)^2 = 13.151$$

then one would choose a part from A. If one wants to maximize the probability that the part's length will be in the interval 998 mm to 1002 mm then

$$p_1 = 0.311$$
$$p_2 = 0.418$$

so that B is favored.

For more complex and wider ranging applications of treatment allocation see Aitchison (1970).

References

Aitchison, J. (1970). Statistical problems of treatment allocation. *Journal of the Royal Statistical Society A* **133**, 206–239.

Geisser, S. (1971). The inferential use of predictive distributions. In *Foundations of Statistical Inference*, V. P. Godambe and D. A. Sprott, (eds.). Toronto: Holt, Rinehart & Winston, 456–469.

Geisser, S., and Eddy, W. F. (1979). A predictive approach to model selection. *Journal of the American Statistical Association* **14**, 153–160.

CHAPTER 6

Perturbation analysis

A Bayesian analysis may depend critically on the modeling assumptions that include prior, likelihood, and loss function. While a loss function is presumably a choice made in the context of particular situations, there is no harm and potentially some gain in investigating the effect on an analysis using alternative loss functions. The likelihood is supposed to represent, to some approximation, the physical process generating the data while the prior reflects subjective views about some of the assumed constructs of this process. Now a likelihood model that has been judged adequate in previous situations similar to a current one is certainly a prime candidate for modeling. It may also have been selected as the most likely when compared with several others and even passed, to a greater or lesser degree, scrutiny by a model criticism technique. However, even in such situations the statistician is still obliged to investigate its present adequacy. A way of addressing this problem is to perturb the "standard" model to a greater or lesser degree in potentially conceivable directions to determine the effect of such alterations on the analysis. While for the strict Bayesian the prior is subjective, it is common knowledge how difficult it often is to subject an investigator or even a statistician to an elicitation procedure that convincingly yields an appropriately subjective prior, even if the parameter is meaningful. As indicated previously, most often the parameter is merely an artifice to promote modeling for prediction. Hence to perturb an investigator's prior or some standard one that appears appropriate is also sensible.

6.1 Types of perturbation

Even when a standard statistical model has proven adequate in data sets similar to a current one at hand, one is obliged to consider the effect of perturbing the standard model in one way or another on the analysis, especially if graphic or other procedures indicate the possibility that the standard model may only be marginally adequate.

There are a large number of possible perturbation schemata. A typically useful one is where $\omega \in \Omega$, an index governing a perturbation schema, is a set of hyperparameters. For $X^{(N)} = (X_1, \ldots, X_N)$, a set of observables, a rather simple example is for $\theta = (\mu, \sigma^2)$,

$$f(x^{(N)}|\theta, \omega) \propto \prod_{j=1}^{N} \left[1 + \frac{(x_j - \mu)^2}{\omega \sigma^2} \right]^{-(\omega+1)/2} \qquad \omega \geq 1$$

where the standard is $\omega \to \infty$, i.e., the normal distribution and the most deviant $\omega = 1$, the Cauchy distribution. Another case is

$$f(x^{(N)}|\theta, \omega) \propto \prod_{j=1}^{N} \frac{k(\omega)}{\sigma} e^{-(1/2)|(x_j - \mu)/\sigma|^{2/(1+\omega)}} \qquad \omega \geq 0$$

where $\omega = 0$ yields the standard normal density and $\omega = 1$ yields the double exponential density.

Further perturbed models that may be used are the Weibull,

$$f(x^{(N)}|\theta, \omega) \propto \prod_{j=1}^{N} \theta \omega x_j^{\omega-1} e^{-\theta x_j^{\omega}} \qquad \omega \geq 1$$

with the standard $\omega_0 = 1$ being the exponential, and the gamma

$$f(x^{(N)}|\theta, \omega) \propto \prod_{j=1}^{N} \theta^{\omega} x_j^{\omega-1} e^{-\theta x_j}$$

also with the standard $\omega = 1$ being the exponential.

A second set is exemplified by a contaminated sampling density

$$f(x^{(N)}|\theta, \omega) \propto \prod_{j=1}^{N} \left[\omega f_1(x_j|\alpha) + (1 - \omega) f_2(x_j|\beta) \right] \qquad 0 \leq \omega \leq 1$$

where say $\omega = 1$ is the standard and α and β are subsets of θ.

Use of ω as an indicator is relevant to situations where ω changes the model distribution to varyingly different but known distributional forms not necessarily in the same family, e.g., say the standard is log normal while the perturbation is gamma. Although this can often be regarded as a special case of either of the first two methods it is best to consider it separately.

A fourth possibility is the use of ω as an exclusion indicator, i.e., $X^{(N)} = (X_1, \ldots, X_N)$ has some standard distributional form under ω_0 but for $\omega \neq \omega_0$ one or more of the X_is have either another distributional form or a completely unspecifiable distribution. In the former case this could mean, for example, that an observation's variance differs from the others or more generally that a parameter set, not under scrutiny, differs for a few of the observations. The latter situation is typically reflected in problems with outliers and aberrant observations that defy satisfactory alternative modeling. Here the perturbation analysis involves what is often termed "an influential case analysis" (Cook and Weisberg, 1982).

A fifth possibility has to do with what one may term periparametric models. Here $\omega = \omega_0$ specifies a standard density while $\omega \neq \omega_0$ specifies all model densities $f(x^{(N)}|\omega)$ that are within a given neighborhood of $f(x^{(N)}|\omega_0)$ determined by varying ω.

A sixth may have to do with possibly inaccurate measurement of the covariates under $\omega \neq \omega_0$ or even the actual responses themselves. All of the above have to do essentially with perturbation of the likelihood.

Similar remarks may be made regarding the prior $p(\theta|\omega)$ and combinations of both likelihood and prior. As a typical example the prior could be a mixture, e.g.,

$$p(\theta|\omega) = \omega p_1(\theta) + (1 - \omega)p_2(\theta)$$

with $\omega = 1$ resulting on the standard $p_1(\theta)$ based on previous information while $p_2(\theta)$ expresses the possibility of another view of the situation. This often results in simpler calculations than having to deal with a likelihood mixture. In particular the use of periparametric perturbation models for additional uncertainty about a "standard" prior seems to be a promising approach especially when combined with a standard likelihood. Here one can examine the extent to which bounds on the "standard" prior can be expanded and still yield robustness for moderate sample size.

6.2　Formal analysis

A formal Bayesian framework for a perturbation analysis for future observables can be delineated (Geisser, 1986, 1988, 1991).

The prediction model considered is, as before, the joint probability function

$$f\left(x^{(N)}, x_{(M)}, \theta \mid \omega\right) = f\left(x_{(M)} \mid x^{(N)}, \theta, \omega\right) f\left(x^{(N)} \mid \theta, \omega\right) p(\theta \mid \omega),$$

$$\omega \in \Omega$$

whence we obtain

$$f\left(x_{(M)} \mid x^{(N)}, \omega\right) = \frac{\int f\left(x^{(N)}, x_{(M)}, \theta \mid \omega\right) d\theta}{\int \int f\left(x^{(N)}, x_{(M)}, \theta \mid \omega\right) d\theta \, dx_{(M)}}.$$

Now assume that $L(a, x_{(M)})$ is the loss incurred in taking action a for a future realization $x_{(M)}$. The average predictive loss

$$\bar{L}_{\omega}(a) = \int L\left(a, x_{(M)}\right) f\left(x_{(M)} \mid x^{(N)}, \omega\right) dx_{(M)}$$

is then minimized for given ω so that

$$\min_{a} \bar{L}_{\omega}(a) = \bar{L}_{\omega}(a_{\omega}^{*})$$

yielding optimal action a_{ω}^{*} when ω is "true." We then consider the difference in the loss when taking action $a_{\omega_0}^{*} = a^{*}$, the optimal action under the standard and when $\omega \neq \omega_0$ is true. We define the differential loss as

$$d(\omega) = \bar{L}_{\omega}(a^{*}) - \bar{L}_{\omega}(a_{\omega}^{*}) \geq 0.$$

One then can examine this loss over a possible range of ω to assess its importance with regard to the action taken under ω_0 and in particular $d^* = \max_{\omega \in \Omega} d(\omega)$. We could also assess its local significance by examining $d(\omega)$ in a neighborhood about ω_0. In fact if ω is a scalar and the second derivative of $d(\omega)$ exists and is continuous the calculation of the curvature at $\omega = \omega_0$, i.e., $d''(\omega_0)$, since $d'(\omega_0) = 0$, could be rather informative regarding local perturbations. For example, a large curvature would indicate that the actions taken could be highly sensitive to a slight variation in the standard model and a serious review of the standard is in order. Little or no curvature would indicate a robustness with respect to a local perturbation in ω.

For a vector ω, the matrix of second derivatives will govern the local curvature and one could assess the maximum curvature, i.e., in the direction of the normed vector associated with the largest root of the Hessian matrix of second derivatives evaluated at the standard $\omega = \omega_0$, say $d''(\omega_0)$.

If one decides that the standard is untenable and the perturbed model reasonable, one possibility is to define a prior distribution for ω and then integrate it out to obtain

$$f\left(x_{(M)}|x^{(N)}\right) = \int p(\omega|x^{(N)})f\left(x_{(M)}|x^{(N)},\omega\right) d\omega.$$

6.3 Other perturbation diagnostics

Often, we are not in a position to discuss decisions or actions that would necessarily flow from a data set and consequently report the predictive distribution itself or some high probability density region for $X_{(M)}$.

For reporting the entire predictive distribution a Kullback–Leibler predictive divergence

$$K(\omega, \omega_0) = E\left[\log f_\omega - \log f_{\omega_0}\right]$$

where $f_\omega = f(x_{(M)}|x^{(N)}, \omega)$ can serve as a reasonable diagnostic. Divergences of this sort were introduced by Johnson and Geisser (1982, 1983) for determining influential observations, one of the particular types of perturbation previously mentioned. In this case the divergences were termed predictive influence functions (PIF).

Predictive diagnostics are most useful in indicating the relative effect of various perturbations.

There may be, however, some difficulty in adequately interpreting globally

$$\max_{\omega \in \Omega} K(\omega, \omega_0)$$

for some of these paradigms.

Another use is to find the direction in which local perturbations have the greatest effect in terms of normal curvature. It can be shown that under suitable regularity conditions that the matrix of second derivatives of $K(\omega, \omega_0)$ for ω a vector of p perturbations, say

$$K''_{\omega = \omega_0} = \mathbf{I}(\omega_0)$$

where $\mathbf{I}(\omega_0)$ is the Fisher Information matrix of the predictive distribution evaluated at $\omega = \omega_0$ (Kullback, 1959). The curvature in direction z, a normed p-dimensional vector where $\omega(t) = \omega_0 + tz$ and $z'z = 1$, is

$$C_z = z'\mathbf{I}(\omega_0)z$$

so that the maximum curvature C^* is in the direction z^*, the vector associated with the maximum root of $\mathbf{I}(\omega_0)$, where

$$C^* = z^{*'}\mathbf{I}(\omega_0)z^*.$$

An examination of the components of z^* will indicate which, namely the larger ones, are those perturbations that relatively most alter the predictive distribution in terms of the divergence (cf. McCulloch, 1989).

Once potentially significant directions are identified, an analysis involving these directions is in order to ascertain whether local departures for them are important enough to vitiate the standard analysis.

The L^1 norm between two densities f and g, favored by Devroye (1987), or the L^2 norm between $\sqrt{f}$ and $\sqrt{g}$, favored by

Pitman (1979) as measures of distance between densities, can also be used here as diagnostics for the predictive distribution. More generally the Hellinger distance between densities raised to the nth power

$$H^n = \int |f^{1/n} - g^{1/n}|^n dx$$

yields these as special cases. For the case here with $n = 2$ we have for predictive densities

$$H^2(\omega, \omega_0) = \int \left(f_\omega^{1/2} - f_{\omega_0}^{1/2} \right)^2 dx_{(M)},$$

which accords with the norm favored by Pitman. Under suitable smoothness conditions, twice the matrix of second derivatives, evaluated at $\omega = \omega_0$ when ω is a vector, is

$$2H_{\omega=\omega_0}^{2''} = \mathbf{I}(\omega_0)$$

again Fisher's Information matrix. Actually there is a whole class of such loss functions that are locally equivalent to $\mathbf{I}(\omega_0)$ (cf. Geisser, 1991). In general, if

$$H(\omega, \omega_0) = E_\omega \phi \left(\frac{f_{\omega_0}}{f_\omega} \right) > \text{ for } \omega \neq \omega_0$$

where ϕ is a convex function and $\phi(1) = 0$ and certain regularity conditions hold then it can be shown that

$$H_{\omega=\omega_0}'' = \phi''(1)\mathbf{I}(\omega_0),$$

based on the work of Burbea and Rao (1982). This is a cogent indicator of the usefulness of this quantity for local perturbation analysis.

The L^1 norm may also be used. While it is unaffected by any one-to-one transformation as is the divergence and H^2, it is analytically awkward and does not discriminate between differences of the two densities when the smaller of the two is large or small, as does H^2 and the divergence.

For the prior mixture case the predictive density of $X_{(M)}$ is

$$f\left(x_{(M)}|x^{(N)}, \omega\right) = \omega f_1\left(x_{(M)}|x^{(N)}\right) + (1-\omega)f_2\left(x_{(M)}|x^{(N)}\right)$$

where

$$f_i\left(x^{(M)}|x_N\right) = \int f\left(x_{(M)}|\theta\right)p_i\left(\theta|x^{(N)}\right) d\theta$$

and

$$p_i\left(\theta|x^{(N)}\right) \alpha f\left(x^{(N)}|\theta\right)p_i(\theta),$$

for $i = 1, 2$. Then we can calculate the information in $f = f(x_{(M)}|x^{(N)}, \omega)$ with respect to ω,

$$I(\omega) = E\left(\frac{\partial \log f}{\partial \omega}\right)^2 = \int \frac{(f_1 - f_2)^2}{(\omega f_1 + (1-\omega)f_2)} dx_{(M)}.$$

Now assuming the existence of

$$\lim_{\omega \to 1} I(\omega) = I(\omega_0) = I(1)$$

and that passage of the limit through the integral sign is valid we obtain

$$I(1) = \int \left(\frac{f_2}{f_1}\right) f_2 \, dx_{(M)} - 1$$

Example 6.1. Let $X_i \sim N(\mu, \sigma^2)$ where σ^2 is known. Let the prior density of μ be a mixture of $N(\tau_i, \eta^2)$ variables, for $i = 1, 2$ where τ_i and η^2 are assumed known.

Then the predictive distribution of X_{N+1} given $x^{(N)}$ is the mixture of $\omega f_1 + (1-\omega)f_2$ where f_i is the density of a

$$N\left(\frac{\eta^2\bar{x} + N^{-1}\sigma^2\tau_i}{\eta^2 + N^{-1}\sigma^2}, \sigma^2 + \sigma^2\eta^2(N\eta^2 + \sigma^2)^{-1}\right)$$

random variable.

Hence we can easily calculate

$$I(1) = e^{\lambda} - 1$$

where

$$\lambda = \frac{\sigma^2 (\tau_1 - \tau_2)^2}{(N\eta^2 + \sigma^2)[\sigma^2 + (N+1)\eta^2]} .$$

For σ^2 unknown we are dealing with "student" t predictive densities for f_1 and f_2, which complicate the calculation for $I(1)$. Now t densities approach normal densities as the degrees of freedom grow and the best normal approximation to a "student" density in terms of minimizing the Kullback divergence is a normal density with the mean and variance of the "student" density. Hence substituting

$$\hat{\sigma}^2 = (N-3)^{-1} \sum (x_i - \bar{x})^2$$

for σ^2 in λ yields, for the σ^2 unknown case with prior $p(\sigma) \propto \sigma^{-1}$,

$$I(1) \doteq e^{\hat{\lambda}} - 1$$

where $\hat{\lambda}(\sigma^2) = \lambda(\hat{\sigma}^2)$. One then must decide how large a curvature is tolerable to admit the use of $\omega = 1$ in the analysis. Note that as N increases the curvature goes to zero, which merely implies the asymptotic irrelevance of the prior.

While the divergence and H^2 are as sensible as any measure of how densities differ overall it is difficult to establish a reasonable calibration so that different values of the divergence or H^2 are easily interpretable, except in a relative sense. Methods for a more suitably direct interpretation that a statistician, and more to the point an investigator, can readily understand can also be defined but they involve rather specific situations. We will present some of these ways of assessing the robustness in terms of predictive regions for $X_{(M)}$. One could restrict oneself to perturbations that could matter as determined locally but we shall retain the same notation as before for two reasons. The first is basically for convenience in that it is possible that the entire ω set may matter and the second is that in certain instances one may not be specifically interested in a local determination. The potential value of the local analysis is the possibility of restricting the dimension of the vector of perturbations to a small set that can more easily be managed by the assessments we now shall propose.

6.4   Predictive region assessments

The first method is to assess the robustness of a $1 - \alpha$ highest probability density region based on the standard ω_0. Suppose this region denoted by $R_{1-\alpha}(\omega_0)$ has volume $V(\omega_0)$ and when perturbed the highest probability density region $R_{1-\alpha}(\omega)$ has volume $V(\omega)$. Let $\nu(\omega)$ be the volume of the intersection of $R_{1-\alpha}(\omega)$ and $R_{1-\alpha}(\omega_0)$ as a function of ω,

$$\nu(\omega) = \text{volume}\left[R_{1-\alpha}(\omega) \cap R_{1-\alpha}(\omega_0)\right]$$

and let

$$\Gamma_\omega = \frac{\nu(\omega)}{M(\omega)},$$

where $M(\omega) = V(\omega)$, $V(\omega_0)$ or $\max[V(\omega), V(\omega_0)]$, be the ratio of the volume of the intersection to the standard, the perturbed or the maximum of the two for the given ω. Then calculate

$$\min_{\omega \in \Omega} \Gamma_\omega = \Gamma_{\omega^*}$$

which now yields the proportion of the region for the "worst" possible case at a given probability $1 - \alpha$. Hence one has an easily interpretable value for assessing the robustness of the data set in terms of a standard analysis involving a $1 - \alpha$ region in the presence of presumably anticipated perturbations.

A second method focuses on the use of the standard region's $R_{1-\alpha}(\omega_0)$ perturbed probability when $\omega \neq \omega_0$. Since

$$\Pr\left[X_{(M)} \in R_{1-\alpha}(\omega_0) \,|\, \omega\right] = \int_{R_{1-\alpha}(\omega_0)} f\left(x_{(M)} | x^{(N)}, \omega\right) dx^{(M)} = 1 - \alpha_\omega$$

we can use

$$\max_{\omega \in \Omega} |1 - \alpha - (1 - \alpha_\omega)| = \max_{\omega \in \Omega} |\alpha_\omega - \alpha|$$

or either of the pair

$$\gamma_1(\alpha) = \max_{\omega \in \Omega} \gamma_1(\alpha, \omega) = \max_{\omega \in \Omega} \frac{|\alpha_\omega - \alpha|}{1 - \alpha_\omega},$$

$$\gamma_2(\alpha) = \max_{\omega \in \Omega} \gamma_2(\alpha, \omega) = \max \frac{|\alpha_\omega - \alpha|}{1 - \alpha},$$

as easily interpretable values. Of possible interest, if no particular value of α is preeminent, is either $\max_\alpha \gamma_1(\alpha)$ or $\max_\alpha \gamma_2(\alpha)$. In addition, sometimes $\gamma_1(1)$ or $\gamma_2(1)$, and $\gamma_1(0)$ or $\gamma_2(0)$ when they exist may be informative. For small changes in ω, $\gamma_1''(\alpha, \omega_0)$ or $\gamma_2''(\alpha, \omega_0)$ for a particular α will be indicative of local behavior.

In summary, the second method is most compelling when some specified region is critical to an analysis, e.g., the effect of the perturbation on the calculation of the probability of an observable exceeding some threshold.

Example 6.2. As a very simple illustration of this, consider $X^{(N)} = (X_1, \ldots, X_N)$ a random sample from

$$f(x|\theta, \omega) = \theta e^{-\theta(x - \omega)} \qquad x \geq \omega \geq 0$$

and noninformative prior

$$p(\theta) \propto \theta^{-1}.$$

Suppose $x_1, \ldots, x_d$ are fully observed realizations and $X_{d+1}, \ldots, X_N$ are independently censored at values $x_{d+1}, \ldots, x_N$. We further suppose, as is almost always the case, that $d \geq 1$ and

$$x_{(1)} = \min(x_1, \ldots, x_d).$$

The predictive distribution function (Geisser, 1982) is then easily calculated to be

$$\Pr\left[X_{N+1} \leq x \mid x^{(N)}, \omega\right] = 1 - \left[1 + \frac{x - \omega}{N(\bar{x} - \omega)}\right]^{-d} \qquad 0 \leq \omega \leq x_{(1)}$$

$$\omega \leq x$$
$$= 0 \qquad \omega > x.$$

Here it is of interest to calculate the probability of a survival threshold, say y,

$$\Pr\left[X_{N+1} > y \mid X^{(N)}, \omega\right] = \left(1 + \frac{y - \omega}{N(\bar{x} - \omega)}\right)^{-d}$$

where the standard say is $\omega = 0$. Of course the divergence and the H^n distances are largely irrelevant for this case but we can easily calculate

$$\max_{0 \leq \omega \leq x_{(1)}} |\alpha_\omega(y) - \alpha(y)| = \left(1 + \frac{y - x_{(1)}}{N(\bar{x} - x_{(1)})}\right)^{-d} - \left(1 + \frac{y}{N\bar{x}}\right)^{-d}$$

for a fixed y or conversely for those values of y such that the quantity on the right is no larger than a given value considered negligible with respect to stating a probability for surviving the threshold.

Numerical illustration of Example 6.2. As an illustration consider the following data reported in Gnedenko et al. (1969, p. 176) consisting of a sample of $N = 100$ items tested and time to failure recorded for each item until 500 standard time units have elapsed. The recorded failure times for 11 items were 31, 49, 90, 135, 161, 249, 323, 353, 383, 436, 477. The remaining 89 items survived the test termination time. If interest is focused on the probability of a future item surviving 500 time units then

$$\Pr\left[X_{N+1} > 500 \mid \omega\right] = \left(1 + \frac{500 - \omega}{47,187 - 100\omega}\right)^{-11}$$

$$= 0.891 \qquad \text{for } \omega = 0$$
$$= 0.890 \qquad \text{for } \omega = x_{(1)} = 31.$$

Hence

$$\max_{0 \le \omega \le 31} |\alpha_{31}(500) - \alpha(500)| = 0.001.$$

On the other hand one might be interested in that value y such that

$$\Pr[X_{N+1} > y | \omega] = 0.5.$$

Here for $\omega = 0$, $y = 3069$ and for $\omega = 31$, $y = 2898$ yielding a maximum relative difference of 5.6%. In passing we also point out here that the maxima for the divergence and the two norms are $K = \infty$, $H^2 \doteq H^1 = 0.01$ and are not particularly informative. The divergence indicates only a difference in support while the norms are approximately and exactly twice the probability assigned to the largest interval over which only one of the densities is supported.

More generally, implementation of these methods in other cases could involve the algebraic or numerical calculation of the intersection of two n-dimensional hyperellipsoids, which could be quite burdensome for $n > 3$. For calculations of this sort, see Chen (1990).

Even more complex situations arise where the highest probability density regions are disconnected. Here one may also want to take into consideration the distance from the standard a perturbed and disconnected region is in ordering the diagnostics discussed above, i.e., not only the size of the nonintersecting disconnected region but its distance in some sense from the standard.

6.5 Predictive influence functions

Much technical and scientific work involves functional relationship of a kind where the sample of observables is stochastically related as

$$Y_j = f(x_j, \beta) + U_j \qquad j = 1, \ldots, N$$

for $x_j' = (x_{1j}, \ldots, x_{qj})$ known values, $\beta' = (\beta_1, \ldots, \beta_p)$ are unknown and U_js are i.i.d. with mean 0 and variance σ^2.

It is often of interest in such situations to determine how an observation or a set of observations influences future predictions. This type of analysis has similarities to the use of an exclusion indicator of Section 3. There the observations were assumed to be identical copies and an observation that yielded appreciable differences in the predictive distributions with and without the observation would in all likelihood be indicative of an aberrant observation. In the case where the observables are independent but not identically distributed as indicated in the functional relationship, appreciable differences in the predictive distributions need not indicate aberrancy but may be the result of the relationship. Hence a pass through the data using a diagnostic to identify highly influential observations is worthwhile. For a thorough exposition on the need for diagnostics to identify influential observations, see Cook and Weisberg (1982).

For a new set of say M observations $Y^{(M)} = (Z_1, Z_2, \ldots, Z_M)$, we assume the same functional and error relationship

$$Z_j = f(x_j, \beta) + U_j \qquad j = 1, \ldots, M$$

where $x_j' = (x_{1j}, \ldots, x_{qj})$.

Now we will make a comparison of

$$f^{(N)} = f(z_1, \ldots, z_M | y^{(N)}) = \int f(z_1, \ldots, z_M | \theta) p(\theta | y^{(N)}) \, d\theta$$

with

$$f^{(N-n)} = f(z_1, \ldots, z_M | y^{(N-n)}) = \int f(z_1, \ldots, z_M | \theta) p(\theta | y^{(N-n)}) \, d\theta$$

where $y^{(N-n)}$ represents a particular partition of the set of observations $y^{(N)} = (y^{(N-n)}, y^{(n)})$ into two sets and $\theta = (\beta, \sigma^2)$.

Of course we may use any of the previously mentioned diagnostics such as the Kullback divergence or the Hellinger distances if there is no particular predictive region of immediate interest. One property of the Hellinger distances is that they are symmetric in the two densities to be compared. The Kullback divergences may be made symmetric by considering the sum of individual divergences

$$K\left(f^{(N)}, f^{(N-n)}\right) + K\left(f^{(N-n)}, f^{(N)}\right).$$

However, we do not believe that this symmetry is necessarily desirable. In fact it may make more sense to consider only the directed divergence

$$K\left(f^{(N-n)}, f^{(N)}\right)$$

as the Predictive Influence Function because we then determine the effect of adding the n observations to the $N-n$ under the tacit assumption that the $N-n$ fit the model.

6.6 Applications to linear regression

One of the most frequently used models is

$$Y_i = x_i'\beta + U_i$$

or in the vector–matrix version as in Section 11 of Chapter 3

$$Y = X\beta + U$$

with

$$U \sim N\left(0, \sigma^2 I_N\right).$$

Further for predicting

$$Z = W\beta + U$$

we have shown that if $p(\beta, \sigma^2) \propto 1/\sigma^2$,

$$Z|y \sim S_M\left[N-p, W\hat{\beta}, s^2 A\right].$$

We shall first deal with an analysis for identifying influential observations by exclusion or, as it is often known, case deletion. If $y = (y_i, y_{(i)})$ is partitioned into y_i the ith observation and $y_{(i)}$ representing the rest then

$$Z|y_{(i)} \sim S_M\left[N - 1 - p, W\hat{\beta}_{(i)}, s^2_{(i)}A_{(i)}\right]$$

where

$$\hat{\beta}_{(i)} = \left(X'_{(i)}X_{(i)}\right)^{-1} X'_{(i)}y_{(i)}$$

and $X_{(i)}$ is X with x'_i, the ith row, deleted and

$$A_{(i)} = I + W\left(X'_{(i)}X_{(i)}\right)^{-1}W'.$$

[This can easily be extended to the partition of y into a set of n deleted and $N - n$ retained (Johnson and Geisser, 1983).]

Hence one needs now to calculate

$$K(f_{(i)}, f) = E\left[\log f_{(i)} - \log f\right] = K(W)$$

where f represents the predictive density of $Z|y$ and $f_{(i)}$ the predictive density of $Z|y_{(i)}$. Although $K(\cdot, \cdot)$ cannot be evaluated explicitly there is an optimal normal density approximation to the student density and it can be used to calculate an appropriate approximation since the divergence between two multivariate normal densities yields an explicit expression. This should be an adequate approximation because a multivariate student density approaches a multivariate normal density asymptotically and the divergence between a "student" density and its normal approximation is minimized for that normal density having the same mean and covariance matrix as the "student" density.

Suppose we wish to predict a single observation at w' a row of W. Hence letting

$$h = w'(X'X)^{-1}w \qquad v_i = x'_i(X'X)^{-1}x_i$$

$$c_i = x'_i(X'X)^{-1}w \qquad t^2_i = \left(y_i - x'_i\hat{\beta}\right)^2 / \left[(N-p)s^2(1 - v_i)\right]$$

then

$$2\hat{K}_i(w)$$

$$= \frac{(N-p-2)t_i^2c_i^2}{(1+h)(1-v_i)}$$

$$+ \frac{(N-p-2)(1-t_i^2)c_i^2}{(N-p-3)(1+h)(1-v_i)} - \log\left[1 + \frac{c_i^2}{(1-v_i)(1+h)}\right]$$

$$+ \frac{N-p-2}{N-p-3}(1-t_i^2) - \log\frac{N-p-2}{N-p-3}(1-t_i^2) - 1.$$

Although this is appropriate if we decide to predict at $W = w'$, in many cases it may not be known where prediction will occur. Hence the use of this as a routine diagnostic is limited. However, in order to circumvent this limitation we can subject W to a probability distribution over a region or more simply set $W = X$, i.e., to simultaneously predict back on the original design matrix X (Johnson and Geisser, 1982, 1983).

Example 6.3. In the case $W = X$, the approximation yields the Predictive Influence Function (PIF) for y_i,

$$2\hat{K}_i = \frac{(N-p-2)v_it_i^2(N-p-4)}{2(1-v_i)(N-p-3)}$$

$$+ \left\{\frac{v_i(N-p-2)}{2(1-v_i)(N-p-3)} - \log\left[1 + \frac{v_i}{2(1-v_i)}\right]\right\}$$

$$+ N\left[\frac{N-p-2}{N-p-3}(1-t_i^2) - \log\frac{N-p-2}{N-p-3}(1-t_i^2) - 1\right].$$

We note that the largest $\hat{K}_i$ yields the most influential observation. From the PIF we can see that the combination of factors that renders an observation influential is essentially its distance from the "center" of the rest of the observations in terms of where it was observed, its inflation of the volume of the predicting ellipsoid, and its lack of fit with respect to the model.

Numerical illustration of Example 6.3. As an illustration of the above we compute $2\hat{K}_i$ for the set of data, Figure 6.1 of 21 observations from a study of cyanotic heart disease in children taken from Mickey et al. (1967). The pertinent data, in compressed form, are presented in Table 6.1, where U^2 and P are as given in Example 4.9. This analysis indicates that case 19 has the largest influence on prediction. Previous case analyses for estimation of the regression coefficient (Andrews and Pregibon, 1978; Dempster and Gasko-Green, 1981) indicated the influence of case 18 and the lack of importance of case 19. However, with regard to prediction, case 19 stands out. While it is true that case 19 has a negligible effect on the regression line itself, the component of $\hat{K}_i$

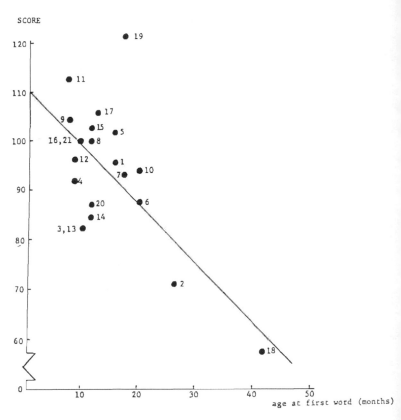

Figure 6.1   *Scatter plot of Gesell data with fitted regression line*

Table 6.1

Case i	$2\hat{K}$	U^2	P
19	2.29	13.03	0.002
18	0.91	0.75	0.409
13	0.10	2.22	0.148

reflecting dispersion of the predictive distribution has a large effect.

Further, the calculation for discordancy for case 19, as given in Example 4.9, is

$$\Pr\left[U^2 > 13.03 | U^2 > 2.22, u_{(C)}\right] = 0.013,$$

which is small enough to cause concern over case 19 as an outlier. A plot of case number vs. P-value is presented in Figure 6.2.

In the previous analysis, no alternative assumption is made about the excluded observation. We now present a simple example where one observation is potentially modeled differently than the others.

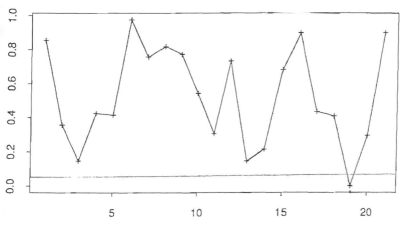

Figure 6.2   *Plot of P-value vs. case number*

Example 6.4. Let $X_1, \ldots, X_N$ be independently distributed as $N(\mu, \sigma^2 | \omega_0)$ under the standard. A possible perturbation is that a single "discordant" value exists (but which one is unknown). Hence, under ω, X_i is $N(\mu, \sigma^2 / \omega | \omega)$.

We assume

$$p(\mu, \sigma^2) \propto \frac{1}{\sigma^2}$$

and calculation of the predictive density under the standard assumption for X_{N+1} yields

$$f(x_{N+1} | x^{(N)}, \omega)$$

$$\propto \left[1 + \frac{(N - 1 + \omega)^2 (x_{N+1} - \bar{u})^2}{(N + \omega)\left[(N - 1)\omega(\bar{x}_{(i)} - x_i)^2 + (N - 1 + \omega)(N - 2)s^2(i)\right]} \right]^{-N/2}$$

where

$$\bar{u} = (N - 1 + \omega)^{-1}\left[(N - 1)\bar{x}_{(i)} + \omega x_i\right]$$

$$(N - 2)s_{(i)}^2 = \sum_{j \neq 1}^{N} (x_i - \bar{x}_{(i)})^2$$

and

$$(N - 1)\bar{x}_{(i)} = \sum_{j \neq i} x_j.$$

The exact calculation of $K_i(\omega)$ can be well approximated by using the optimal normal approximation to the student density. Hence

$$2\hat{K}_i(\omega) \doteq \frac{(\bar{x} - \bar{u})^2 S}{V^2} + \frac{S}{V} - \log \frac{S}{V} - 1$$

where

$$S = \frac{N + 1}{(N - 3)N} \sum_{j=1}^{N} (x_j - \bar{x})^2$$

$$V = \frac{N + \omega}{N - 3} \left[\frac{(N - 1)\omega(\bar{x}_{(i)} - x_i)^2}{(N - 1 + \omega)^2} + \frac{(N - 2)s_{(i)}^2}{N - 1 + \omega} \right].$$

Table 6.2 *Weight of Eggs in Grams*

51.52	49.73	60.60	50.12	58.51	59.25	55.87	58.74	53.00	54.00 57.88 59.83
53.37	65.07	50.11	30.24	56.06	56.11	56.88	56.83	51.55	55.04 50.43 48.22
52.61	55.14	54.84	56.07	62.20	60.58	55.66	54.90	60.04	56.23 52.18 55.06
52.74	59.59	53.45	54.57	56.28	46.06	59.20	55.61	53.04	56.65 57.89 52.54
58.28	49.06								

Numerical illustration of Example 6.4. A random sample of 50 eggs, one from each of 50 different chickens, was obtained. The weight of each egg is given in grams (Table 6.2). It is assumed that at most one of the eggs was generated from a population with a variance different from the rest. The value of $2\hat{K}(\omega)$ for the next egg obtained, assuming it to have been laid by a chicken with standard variance, is calculated for three sample egg weights—30.24, 46.06, and 55.14. The first appears to be furthest from the sample mean, the next one the second furthest, and the last close to the sample mean. It is clear from Figure 6.3 that the effect of egg weight 30.24 on the predictive distribution is substantially greater than any other observation as ω varies and obviously has high curvature in the neighborhood of $\omega = 1$. For egg weight 46.06, $2\hat{K}(\omega)$ changes very slowly while for egg weight 55.14 there is virtually no change in $2\hat{K}(\omega)$ so that any other observation will minimally effect the predictive distribution for a small change in its variance from the standard.

A hyperparameter perturbation where the variances of all the observations are locally perturbed from the standard variance is also a possible model perturbation. An analysis of this kind is now presented. For the previous regression model it is assumed that

$$U \sim N(0, \sigma^2 R^{-1})$$

where

$$R = \begin{pmatrix} 1 + \omega_1 & & 0 \\ & \ddots & \\ 0 & & 1 + \omega_N \end{pmatrix}.$$

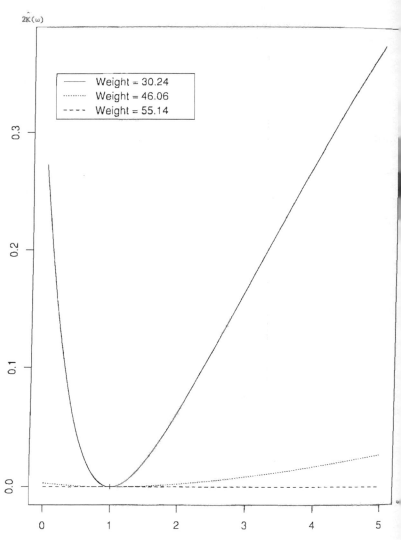

Figure 6.3   *Plot of* $2\hat{K}(\omega)$ *vs.* ω *for three egg weights*

By transforming $X_\omega = R^{1/2}X$ and $Y_\omega = R^{1/2}Y$ then

$$Y_\omega = X_\omega\beta + U.$$

For a future $Y_{N+1} = w'\beta + u_{N+1}$ where $u_{N+1} \sim N(0, \sigma^2)$ we obtain its predictive distribution to be

$$S\left[N - p, w'\beta_\omega, s_\omega^2(1 + h_\omega)(N - p)\right]$$

where

$$h_\omega = w'(X'RX)^{-1}w$$
$$\hat\beta_\omega = (X'RX)^{-1}X'Ry$$
$$s_\omega^2 = y'\left[R - RX(X'RX)^{-1}X'R\right]y/(N - p).$$

Then it can be shown (Lavine, 1987) that the curvature in direction z where $z'z = 1$ is

$$C_z = z'\left[B_1 + B_2 + B_3 + B_4\right]z$$

where

$$B_1 = \frac{(N - p)}{(N - p + 3)2(1 + h)^2}(m \circ m)(m \circ m)'$$

$$B_2 = \frac{-1}{(1 + h)s^2}(m \circ m)(r \circ r)$$

$$B_3 = \frac{1}{2(N - p - 3)(N - p)s^4}(r \circ r)(r \circ r)'$$

$$B_4 = \frac{N - p + 1}{(N - p + 3)(1 + h)s^2}(r \circ m)(r \circ m)'$$

for s_ω^2 and h_ω being s^2 and h when $R = I$ and $\circ$ denotes elementwise multiplication for the vectors

$$r = \left[I - X(X'X)^{-1}X'\right]y$$

and

$$m = X(X'X)^{-1}w.$$

The four $p \times p$ matrices are all of rank 1 so that their sum is of most rank 4. The maximum curvature then is in the direction z^* the vector associated with the maximum root of $B = B_1 + B_2 + B_3 + B_4$. The coordinates of

$$z^* = (z_1^*, \ldots, z_N^*)'$$

represent a worst case scenario that depends of course on the given w. These can now be plotted against a covariate, i.e., a column of X. Large values of z_i^* could indicate a local sensitivity to heteroscedasticity as a function of that covariate when predicting Y_{N+1} at the given w. These plots could vary substantially depending on the column and the given w. For general W or simultaneous prediction of several future observables, the formulas are more complex with only a slight simplification for the back prediction case when $W = X$. The latter is of interest when an overall assessment of the effect on prediction is required in the presence of slightly perturbed variances.

References

Andrews, D. F., and Pregibon, D. (1978). Finding outliers that matter. *Journal of the Royal Statistical Society B* **40**, 85–93.

Burbea, J., and Rao, C. R. (1982). Entropy differential metric, distance and divergence measures in probability spaces: A unified approach. *Journal of Multivariate Analysis* **12**, 575–596.

Chen, C. N. (1990). New diagnostic measures in the linear model. Unpublished doctoral dissertation. University of Minnesota.

Cook, R. D., and Weisberg, S. (1982). *Residuals and Influence in Regression*. New York: Chapman and Hall.

Dempster, A. P., and Gasko-Green, M. (1981). New tools for residual analysis. *Annals of Statistics* **9**, 945–959.

Devroye, L. (1987). *A Course in Density Estimation*. Boston: Birkhauser.

Geisser, S. (1982). Aspects of the predictive and estimative approaches in the determination of probabilities. *Biometrics* **38**, suppl., 75–93.

Geisser, S. (1986). Contribution to discussion. *Journal of the Royal Statistical Society B* **47**, 31–32.

Geisser, S. (1988). The future of Statistics in retrospect. In *Bayesian Statistics 3*, J. M. Bernardo et al. (eds.). Oxford: Oxford University Press, 147–158.

Geisser, S. (1991). Diagnostics, divergences, and perturbation analysis. In *Directions in Robust Statistics and Diagnostics*, W. Stahel and S. Weisberg (eds.). Berlin: Springer-Verlag, 89–100.

Geisser, S. (1992a). Bayesian perturbation diagnostics and robustness. *Bayesian Analysis in Statistics and Econometrics, Lecture Notes in Statistics* 75. Berlin: Springer-Verlag, 289–302.

Gnedenko, B. B., Belyayev, Y. K., and Solovyev, A. D. (1969). *Mathematical Methods of Reliability Theory*. New York: Academic Press.

Johnson, W., and Geisser, S. (1982). Assessing the predictive influence of observations. In *Statistics and Probability Essays in Honor of C. R. Rao*, G. Kallianpur, P. R. Krishnaiah, and J. K. Ghosh (eds.). Amsterdam: North-Holland, 343–348.

Johnson, W., and Geisser, S. (1983). A predictive view of the detection and characterization of influential observations in regression analysis. *Journal of the American Statistical Association* **78**, 137–144.

Kullback, S. (1959). *Information Theory and Statistics*. New York: John Wiley.

Lavine, M. (1987). Prior influence in Bayesian statistics. Unpublished doctoral dissertation, University of Minnesota.

McCulloch, R. (1989). Local model influence. *Journal of the American Statistical Association* **84**, 472–478.

Mickey, M. R., Dunn, O. J., and Clark, V. (1967). Note on the use of stepwise regression in detecting outliers. *Biomedical Research* **1**, 105–109.

Pitman, E. J. G. (1979). *Some Basic Theory for Statistical Influence*. London: Chapman and Hall.

CHAPTER 7

Process control and optimization

Suppose we have performed a series of trials wherein a trial consists of knowing or assigning a set of values x_i in an assumed functional relationship

$$Y_i = g(x_i, \beta) + U_i \qquad i = 1, \ldots, N$$

which depends on unknown β and error term U_i having mean 0 and variance σ^2.

Our objective will be to obtain a future outcome

$$Z = g(x_{N+1}, \beta) + U_{N+1}$$

to be as close as possible, in some to be defined sense, to a value y_0 given that we can control the set of values x_{N+1}.

7.1 Control methods

The methods imply that we calculate the predictive probability function

$$f(z|x_{N+1}, y^{(N)})$$

and then choose a particular loss function that will suit our control purposes for determining the assigned value $\hat{x}_{N+1}$. For example, we may choose that optimal x_{N+1} such that it satisfies

$$\min_{x_{N+1}} E\left[(Z - y_0)^2 | x_{N+1}\right]$$

or

$$\max_{x_{N+1}} \Pr\left[Z \in R(y_0)|x_{N+1}\right]$$

where $R(y_0)$ is a region of interest located about y_0. In the latter case, as the region collapses about the point y_0, we maximize the predictive density at $z = y_0$ with respect to x_{N+1}. Optimization may require other attributes of the response then being as close as possible to some known value. For example, we may want a maximum or minimum response for a function of z subject to constraints on a function of x_{N+1}. When the U_is are independent copies, the optimal control problem for a sequence of $Z_1, \ldots, Z_M$ could be handled by backward induction (Bellman, 1957), when M is known. The calculations however can be quite arduous. When M is unknown one could just sequentially update the predictive probability function so that

$$f\left(z_j|y^{(N)}, x^{(N)}, \ldots, z^{(j-1)}, \hat{x}^{(j-1)}, x_{N+j}\right)$$

is operated on for control of Z_j where $x^{(N)}, z^{(j-1)}, \hat{x}^{(j-1)}$ represent the sets $(x_1, \ldots, x_N)$, $(z_1, \ldots, z_{j-1})$, and $(x_{N+1}, \ldots, x_{N+j-1})$, respectively. Even for M known, this generally should not give very different control values than backward induction. Further, the additional loss incurred may be quite negligible for most loss functions. When $U_1, \ldots, U_N, U_{N+1}, \ldots, U_{N+M}$ are dependent there is much greater complexity in making the appropriate calculations.

7.2   Poisson regulation

A simple case involving particle emission is given by Dunsmore (1969). A substance containing a known amount x of radioactive material emits Y radioactive particles per unit time according to the Poisson law for Y with mean $x\theta$, i.e.,

$$\Pr[Y = y|x, \theta] = \frac{e^{-x\theta}(\theta x)^y}{y!} \qquad y = 0, 1, \ldots.$$

Suppose we have N such independent emissions $Y_1, \ldots, Y_N$ with associated known amounts $x_1, \ldots, x_N$. We wish to regulate a future outcome Z to be close to y_0 by choice of x_{N+1}. Assume $p(\theta) \propto \theta^{-1}$, then for the predictive probability function of Z we obtain

$$\Pr\left[Z = z | x_{N+1}, y^{(N)}, x^{(N)}\right]$$

$$= \binom{t+z-1}{t-1}\left(\frac{x_{N+1}}{u+x_{N+1}}\right)^z \left(\frac{u}{u+x_{N+1}}\right)^t \qquad t \geq 1$$

where $t = \sum_{i=1}^{N} y_i$, $u = \sum_{i=1}^{N} x_i$. Hence Z has the negative binomial distribution.

For quadratic loss we need to minimize

$$E\left[(Z - y_0)^2 | x_{N+1}\right] = V(Z) + \left[E(Z) - y_0\right]^2$$

$$= \frac{x_{N+1} t (u + x_{N+1})}{u^2} + \left(\frac{t x_{N+1}}{u} - y_0\right)^2$$

with respect to x_{N+1}. Minimizing the above yields

$$\tilde{x}_{N+1} = \frac{u\left[y_0 - (1/2)\right]}{(t+1)}, \qquad \text{if } y_0 > 1/2$$

$$\tilde{x}_{N+1} \equiv 0, \qquad \text{if } y_0 \leq 1/2.$$

Maximization of

$$\Pr\left[Z = y_0 | x_{N+1}, y^{(N)}, x^{(N)}\right]$$

with respect to x_{N+1} yields

$$\hat{x}_{N+1} = \frac{y_0 u}{t}.$$

Thus

$$\tilde{x}_{N+1} = \hat{x}_{N+1} = 0 \qquad \text{for } y_0 = 0$$

$$\hat{x}_{N+1} < \tilde{x}_{N+1} \qquad \text{for } 0 < y_0 \leq 1/2$$

$$\tilde{x}_{N+1} = t\hat{x}_{N+1}\left[\frac{1 - (1/2y_0)}{t+1}\right] \qquad \text{for } y_0 > 1/2.$$

Hence for $y_0 > 0$, $\tilde{x}_{N+1} < \hat{x}_{N+1}$ and they can be quite different for small y_0. For example if $y_0 = 1$ then $\tilde{x}_{N+1} < (1/2)\hat{x}_{N+1}$. As y_0 increases they tend to approach one another.

To maximize

$$\Pr\left[\, y_0 - a \le Z \le y_0 + b \,|\, x_{N+1}\right]$$

with respect to x_{N+1}, we set its derivative to 0. This will yield a polynomial in x_{N+1} which can be solved numerically for the appropriate root which maximizes this probability.

Sometimes we require a two-step solution where we are controlling for future values Z_1 and Z_2 by regulating x_{N+1} and x_{N+2}. In this case one solution requires setting $z_1 = y_{01}$ and $z = y_{02}$ in the joint probability function.

$$\Pr\left(Z_1 = y_{01}, Z_2 = y_{02} \,|\, x_{N+2}, x_{N+1}, y^{(N)}, x^{(N)}\right) = f \qquad (7.1)$$

and then maximizing (7.1) w.r.t. to x_{N+1} and x_{N+2}. Now setting

$$\frac{\partial \log f}{\partial x_{N+2}} = \frac{y_{02}}{x_{N+2}} - \frac{t + y_{01} + y_{02}}{u + x_{N+1} + x_{N+2}} = 0$$

and

$$\frac{\partial \log f}{\partial x_{N+1}} = \frac{y_{01}}{x_{N+1}} - \frac{t + y_{01} + y_{02}}{u + x_{N+1} + x_{N+2}} = 0$$

and solving the above simultaneously yields

$$\hat{x}_{N+2} = \frac{y_{02}u}{t}$$

$$\hat{x}_{N+1} = \frac{y_{01}u}{t}.$$

If we consider M-steps ahead, it is clear that maximizing the probability $Z_i = y_{0i}$, $i = 1, \ldots, M$ results in

$$\hat{x}_{N+i} = \frac{y_{0i}u}{t} \qquad i = 1, \ldots, M.$$

This solution differs from a sequential updating procedure where it is easy to show that at the $N + i$ step the solution is

$$\hat{X}_{N+i} = \frac{y_{0i}(u + x_{N+1} + \cdots + x_{N+i-1})}{t + y_{N+1} + \cdots + y_{N+i-1}},$$

by separately maximizing each conditional predictive probability function

$$\Pr\left[Z_j = y_{0j} | z_{j-1} = y_{N+j-1}, x_{N+j}, x^{(N+j-1)}, y^{(N+j-1)}\right].$$

In terms of loss functions the first solution maximizes the chance that all $Z_1, \ldots, Z_M$ will be on target, and consequently the entire loss is endured even if one Z_i is not on target. The second solution treats the losses individually in that a loss is sustained only for those not on target. Other situations, where a loss is suffered only if k or more out of the M are not on target, can also be considered. Of course the solution one would use would depend on the criterion specified.

In regard to minimization of a squared loss of the type

$$\sum_{i=1}^{M} E\left[(Z_i - y_{0i})^2 | x_{N+1}, \ldots, x_{N+M}\right],$$

an explicit solution for the M-step solution is not available. However, the sequential updating procedure here is simply

$$\bar{x}_{N+i} = \frac{(u + x_{N+1} + \cdots + x_{N+i-1})[y_{0i} - (1/2)]}{t + y_{N+1} + \cdots + y_{N+i-1} + 1} \qquad i = 1, 2, \ldots$$

7.3   Control in linear regression

The next paradigm involves linear regression. Using the setup of Section 11 of Chapter 3, the predictive density of a single future observation that is modeled as

$$Y_{N+1} = x'_{N+1}\beta + U_{N+1}$$

for $U_{N+1} \sim N(0, \sigma^2)$, is

$$f(z|y) = \left\{ \Gamma[(N+1-p)/2] \right.$$

$$\times \left[1 + \frac{\left(z - x'_{N+1}\hat{\beta}\right)^2}{(N-p)s^2\left(1 + x'_{N+1}(X'X)^{-1}x_{N+1}\right)} \right]^{-(N+1-p)/2} \right\}$$

$$\div \left\{ \pi^{1/2}\Gamma[(N-p)/2](N-p)^{1/2} \right.$$

$$\left. \times \left[s^2\left(1 + x'_{N+1}(X'X)^{-1}x_{N+1}\right) \right]^{1/2} \right\}$$

where $\hat{\beta} = (X'X)^{-1}X'y$. For the squared error loss function

$$E(Z - y_0)^2 = V(Z) + \left(y_0 - x'_{N+1}\hat{\beta} \right)^2$$

$$= \frac{N-p}{N-p-2}\left(1 + x'_{N+1}(X'X)^{-1}x_{N+1}\right)s^2$$

$$+ \left(y_0 - x'_{N+1}\hat{\beta} \right)^2.$$

Setting the derivatives with respect to the components of x_{N+1} to zero yields as solution

$$\dot{x}_{N+1} = y_0\hat{\beta}\left(\frac{N-p}{N-p-2}s^2(X'X)^{-1} + \hat{\beta}\hat{\beta}' \right)^{-1}.$$

This result was obtained by Zellner and Chetty (1965).

Let $X = (e, X_2)$ have ones in the first column and $x'_{N+1} = (1, w')$. In this case

$$E(Z - y_0)^2 = \frac{N-p}{N-p-2}\left[1 + (1, w')(X'X)^{-1}\binom{1}{w} \right]s^2$$

$$+ \left(y_0 - \hat{\beta}_1 - w'\hat{\beta}_{(1)} \right)^2$$

$$= \frac{N-p}{N-p-2}\left[1 + \frac{1}{N} + (w - \bar{x})'A^{-1}(w - \bar{x}) \right]s^2$$

$$+ \left(y_0 - \hat{\beta}_1 - w'\hat{\beta}_{(1)} \right)^2$$

where $N\bar{x}' = e'X_2$, $A = X'_2[I - (ee'/N)]X_2$, and $\hat{\beta} = (\hat{\beta}_1, \hat{\beta}_{(1)})$.

Setting the derivative of $E(Z - y_0)^2$ with respect to w to zero yields the minimum

$$\tilde{w} = \left[(y_0 - \hat{\beta}_1)\hat{\beta}_{(1)} + \frac{N-p}{N-p-2}s^2A^{-1}\bar{x} \right]$$
$$\times \left(\frac{N-p}{N-p-2}s^2A^{-1} + \hat{\beta}_{(1)}\hat{\beta}'_{(1)} \right)^{-1} .$$

Setting $p = 2$ yields the solution given by Aitchison and Dunsmore (1975).

The solution, for the loss function that maximizes $f(y_0|y^{(N)})$, would require a numerical search. Such a search could begin in the region about either $\hat{x}_{N+1}$ or $\hat{w}$ depending on the model. More generally

$$\Pr\left[y_0 - a \le Z \le y_0 + b | x_{N+1}, y^{(N)}, X \right]$$

can be maximized with respect to x_{N+1} or w numerically.

For complications involving cost in selecting controlling covariates and selecting those that optimally regulate the predicted value with respect to minimizing cost (see Lindley, 1968). In particular, a solution has been given for maximizing $\Pr(Z > z|x_{N+1}, y^{(N)}, X)$ subject to $C'x_{N+1} \le k$ for $C' = (C_1, \ldots, C_p)$, $C_i \ge 0$, $k > 0$ and all the components of x_{N+1} are nonnegative (Cain and Owen, 1990).

7.4   Partial control

Situations arise where some of the values of w cannot be controlled but are given. Let

$$w' = (t', u')$$

where $t' = (t_1, \ldots, t_r)$ is given and $u' = (u_{r+1}, \ldots, u_{p-1})$ controllable. Let

$$X_2 = (X_3, X_4)$$

where X_3 represents the first r columns of X_2 and X_4 the last p-r-1 columns, respectively, and

$$N\bar{x}' = e'X_2 = (e'X_3, e'X_4) = N(\bar{x}^{(1)\prime}, \bar{x}^{(2)\prime}).$$

Hence

$$E(Z - y_0)^2 = \frac{s^2(N-p)}{(N-p-2)}\left[1 + \frac{1}{N} + (t - \bar{x}^{(1)})' A^{11}(t - \bar{x}^{(1)}) \right.$$
$$+ 2(u - \bar{x}^{(2)})' A^{21}(t - \bar{x}^{(1)})$$
$$\left. + (u - \bar{x}^{(2)})' A^{22}(u - \bar{x}^{(2)}) \right]$$
$$+ \left(y_0 - \hat{\beta}_1 - t'\hat{\gamma}_1 - u'\hat{\gamma}_2 \right)^2$$

where

$$A^{-1} = \begin{pmatrix} A^{11} & A^{12} \\ A^{21} & A^{22} \end{pmatrix}$$

and

$$\hat{\beta}'_{(1)} = (\hat{\gamma}'_1, \hat{\gamma}'_2).$$

Then minimization of $E(Z - y_0)^2$ yields

$$\bar{u} = \left[\hat{\gamma}_2 \hat{\gamma}'_2 + \frac{s^2(N-p)}{N-p-2} A^{22} \right]^{-1}$$
$$\times \left\{ \frac{s^2(N-p)}{N-p-2}\left[A^{21}(t - \bar{x}^{(1)}) + A^{22}\bar{x}^{(2)} \right] + \left(y_0 - \hat{\beta} - t'\hat{\gamma}_1 \right)\hat{\gamma}_2 \right\}.$$

See Zellner (1971) for other variations.

7.5 Autoregressive regulation

In many situations the generator of the observations may be a single unit, perhaps a machine or individual. For example, severe diabetics need to control their glucose level by carefully controlled infusions of insulin. This must be done at least every day and in some cases several times a day. The amount of insulin required to regulate the glucose level depends on an individual's current level and possibly other variables that are uncontrolled but known such as diet, weight, and other concomitants. It may be possible to approximately model situations such as this as

$$Y_{t+1} = \alpha y_t + \gamma v_t + \beta' x_t + U_t, \qquad U_t \sim N(0, \sigma^2) \qquad t = 1, \ldots, N$$

where y_t is the glucose level at time t, v_t is the amount of insulin infused at time t, and x_t is the set of other uncontrolled but known values. Of course Y_{t+1} should also depend on the total amount of insulin at time t, i.e., that amount already that is in the body plus v_t, but we are assuming that the amount in the body is not easily measurable and that Y_t itself reflects to a great degree the amount of insulin already there. The problem here is to obtain a realized value of Y_{t+1} that is as close as possible to y_0, knowing $Y_t = y_t$ and the outcome x_t by assigning a value to v_t. Clearly this paradigm could also fit other chronic conditions requiring periodic treatment for their regulation such as blood pressure, cholesterol level, and arthritis. This type of modeling may also be appropriate for some economic and engineering data series.

For the squared error loss function it can be shown that the solution for v_t, having observed $(y_1, \ldots, y_t) = y^{(t)}$ and assigning a joint prior distribution to α, γ, β, and σ^2 is

$$\tilde{v}_t = \left[y_0 E(\gamma | y^{(t)}) - y_t E(\gamma \alpha | y^{(t)}) - E(\gamma \beta' | y^{(t)}) x_t \right] \left[E(\gamma^2 | y^{(t)}) \right]^{-1}$$

when the expectation is over the posterior distribution of the parameters. The second loss function requires maximization of the predictive density of Y_{t+1} evaluated at $Y_{t+1} = y_0$ with respect to v_t, namely

$$\max_{v_t} f_{Y_{t+1}}(y_0 | y^{(t)}, x^{(t)}) = \max_{v_t} E\left[f_{Y_{t+1}}(y_0 | x^{(t)}, \alpha, \beta, \gamma, \sigma^2) \right]$$

with the expectation over the posterior distribution of the parameters. The predictive density of Y_{t+1}, even under the simplest of prior densities on the parameters, is rather complicated. Hence it may be easier to obtain the result by interchanging derivatives and integral signs as indicated by the quantity on the right-hand side above. Other loss functions that could be of importance here are asymmetric ones, e.g., linear with different slopes on either side of y_0 or quadratic on one side and linear on the other side. In particular, one that is essentially exponential on one side and linear on the other could be of particular value. For example, certain diabetics monitored daily and susceptible to large variation in glucose level may need much more "protection" against having too much insulin than having too little to protect against insulin shock. Here we would need to minimize

$$E\left[e^{b(y_0 - Y_{t+1})} - b(y_0 - Y_{t+1}) - 1 \right]$$

with respect to v_t in the predictive density of Y_{t+1}. This has been termed the LINEX loss function by Varian (1975) and a solution provided by Zellner (1986) in a standard multiple regression situation with known variance. The solution for the case derived here would require calculating numerically that v_t^* that satisfies

$$E\left[\gamma e^{b(v_0 - \alpha y_t - \gamma v_t - \beta' x_t) + (b^2 \sigma^2)/2} | y^{(t)} \right] = E\left[\gamma | y^{(t)} \right]$$

where the expectation is over the joint posterior distribution of α, β, and γ, for known σ^2. When σ^2 is unknown the solution may or may not exist depending on the posterior distribution of σ^2. Because prolonged periods of too little insulin could lead to acidosis, the linex method may use $Y_{t+1} - y_0$ instead of $y_0 - Y_{t+1}$ to accommodate this possibility. In fact, depending on the history of the patient's problems, the linex or some other asymmetric loss function could be varied in either direction during the course of treatment until some stability is achieved. Once stability is reached one of the other loss functions might be used. For the control of a cholesterol level exponential loss for values above some interval, constant in the interval, and linear below, may be appropriate. Depending also on what agent is being used to regulate the response, more than likely, constraints on the allowed amount of the agent to prevent other deleterious effects would also be in order.

The solutions for the various loss functions will depend on the prior densities assigned to the set of parameters α, β, γ, and σ^2. However, since there will be, in a rather short time, considerable data, the effect of using a "noninformative" prior say,

$$p(\alpha, \beta, \gamma, \sigma^2) \propto \frac{1}{\sigma^2},$$

for $\beta \in R^p$, $\gamma_0 \leq \gamma \leq \gamma_1$, $\sigma^2 > 0$, and $0 < \alpha < 1$, may be of negligible consequence to the regulation procedure. This will certainly be the case where previous data on the patient were at hand that are then combined with the prior density given above. Note also we have only considered the sequential updating procedure, which is difficult enough for which to obtain solutions. If we knew the number of future values for t, for which this process would be validly used, one could attempt an optimal backward induction solution. However, since (1) the horizon is generally unknown, (2) the model itself may need alteration as the process continues, (3) the complexity of a solution to a very large (possibly daily for the lifetime of a diabetic, excluding model changes) horizon, and (4) the difference between the sequential updating and the backward induction is in all likelihood negligible, then sequential updating should suffice. However differing loss functions could result in quite different control values.

7.6   Illustration of autoregressive regulation

Monthly data on income and advertising expenditures of a corporation were collected for 36 months. For the first 30 months advertising was conducted using medium A and for the last 6 months by medium B. The data are given in Table 7.1. For the next month ($t = 37$) the advertising budget was cut back to \$3600 $= x_{37}$ to be spent on either A or B. In order to decide which medium to use to maximize income, the advertising department used the following model

$$Y_{t+1} = \alpha y_t + \gamma v_t + \beta_0 + \beta_1 x_t + U_t$$

where U_t were independently distributed as $N(0, \sigma^2)$, $\alpha \in (0, 1)$,

Table 7.1 *Advertising Expenditures X and Sales Income Y in Hundreds of Dollars*

t	x_t	y_{t+1}	t	x_t	y_{t+1}	t	x_t	y_{t+1}
1	52	4435	13	35	2969	25	41	2520
2	54	4295	14	35	3320	26	42	2809
3	57	4364	15	36	3211	27	46	3190
4	55	4028	16	38	2660	28	58	3528
5	51	3734	17	37	2824	29	61	3729
6	52	4140	18	31	2396	30	64	3393
7	49	3863	19	39	2203	31	59	3233
8	47	3677	20	34	2392	32	55	3492
9	44	3651	21	31	2770	33	54	3244
10	41	4078	22	30	2643	34	58	3218
11	38	4116	23	35	1862	35	56	3541
12	41	3189	24	38	1938	36	53	3821

$(\gamma, \beta_0, \beta_1) \in R^3$

$$v_t = \begin{cases} 1 & \text{for medium } A \\ -1 & \text{for medium } B. \end{cases}$$

It was assumed that all the parameters were a priori independently distributed such that

$$p(\sigma^2, \alpha, \gamma, \beta_0, \beta_1) \propto \frac{1}{\sigma^2}$$

for $\sigma^2 > 0$, $\alpha \in (0,1)$, and $\tau = (\gamma, \beta_0, \beta_1) \in R^3$, and initial value $y_1 = 465,000$. Hence the predictive density of Y_{38} is given by

$$f\left(y_{38} | y^{(37)}, x^{(36)}, x_{37}\right)$$

$$= \int f\left(y_{38} | \alpha, \tau, \sigma^2, y_{37}, x_{37}\right) p\left(\alpha, \tau, \sigma^2 | y^{(37)}, x^{(36)}\right) d\alpha \, d\tau \, d\sigma^2$$

$$\doteq \frac{1}{M} \sum_{i=1}^{M} f\left(y_{38} | \alpha_i, \tau_i, \sigma_i^2, y_{37}, x^{(36)}, x_{37}\right)$$

where α_i, τ_i, and σ_i^2 are random draws from the posterior density

$$p\left(\alpha, \tau, \sigma^2 | y^{(37)}, x^{(36)}\right)$$

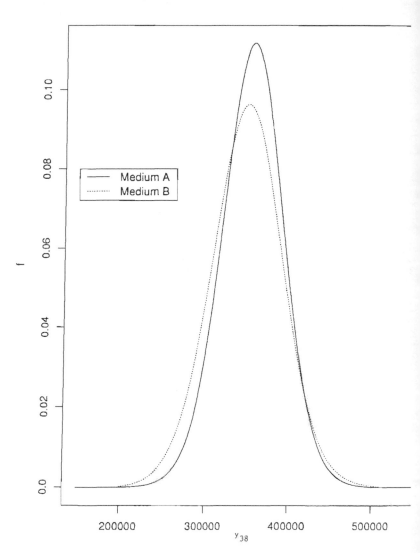

Figure 7.1   *Predictive densities of Y_{38} at $x_{37} = 3600$ for A and B*

and $y^{(37)} = (y_1, \ldots, y_{37})$, $x^{(36)} = (x_1, \ldots, x_{36})$. The draws from the posterior density are obtained via the Gibbs sampler (Gelfand and Smith, 1990; Casella and George, 1992), and then used in the known sampling density of Y_{38} that is $N(\alpha y_{37} + \gamma v_{37} + \beta_0 + \beta_1 x_{37}, \sigma^2)$ for a large set of values of y_{38}. A plot of the predictive density of Y_{38} in Figure 7.1 is then obtained for both A and B for $x_{37} = 3600$ over the range of values that contain almost all of the density. The predictive means and standard deviations of Y_{38} under A and B are also calculated in a similar manner. Hence we obtain

$$E(Y_{38}|A) \doteq 355{,}300 \qquad \sigma(Y_{38}|A) \doteq 3690$$
$$E(Y_{38}|B) \doteq 349{,}300 \qquad \sigma(Y_{38}|A) \doteq 4290.$$

This favors A slightly in terms of the predictive means. Also from the graph in Figure 7.1, it is clear that most of the density of A is to the right of B. Calculation discloses that up to about $y_{38} \doteq$ 393,000 the predictive distribution of Y_{38} under B is larger than under A. For y_{38} beyond 393,000 the situation is reversed. However, the maximum difference in probability calculated numerically in the former case is about .08 while in the latter it is about 0.01. In summary,

$$0 \le \Pr[Y_{38} \ge y_{38}|A] - \Pr[Y_{38} \ge y_{38}|B] \le 0.08 \quad \text{for } y_{38} \le 393{,}000$$
$$0 \le \Pr[Y_{38} \ge y_{38}|B] - \Pr[Y_{38} \ge y_{38}|A] \le 0.01 \quad \text{for } y_{38} > 393{,}000.$$

Hence, if the choice is based on the greater probability that $Y_{38} \ge y_{38}$, A would be chosen if $y_{38} \le 393{,}000$ and B otherwise, but since the maximum that each of these probabilities attains is negligible, there is little to choose between the media, if the current level of sales is to be maintained.

References

Aitchison, J., and Dunsmore, I. R. (1975). *Statistical Prediction Analysis.* Cambridge: Cambridge University Press.

Bellman, R. (1975). *Dynamic Programming.* Princeton, NJ: Princeton University Press.

Cain, M., and Owen, R. J. (1990). Regression levels for Bayesian predictive response. *Journal of the American Statistical Association* **85** (409), 228–231.

Casella, G., and George, E. I. (1992). Explaining the Gibbs sampler. *The American Statistician* **46** (3), 167–174.

Dunsmore, I. R. (1969). Regulation and optimization. *Journal of the Royal Statistical Society B* **31**, 160–170.

Gelfand, A. E., and Smith, A. F. M. (1990). Sampling based approaches to calculating marginal densities. *Journal of the American Statistical Association* **85**, 398–409.

Lindley, D. V. (1968). The choice of variables in multiple regression. *Journal of the Royal Statistical Society B* **30**, 31–66.

Varian, H. R. (1975). A Bayesian approach to real estate assessment. In *Studies in Bayesian Econometrics and Statistics*, S. Fienberg and A. Zellner (eds.). Amsterdam: North-Holland, 195–208.

Zellner, A. (1971). *An Introduction to Bayesian Inference Econometrics.* New York: John Wiley.

Zellner, A. (1986). Bayesian estimation and prediction using asymmetric loss functions. *Journal of the American Statistical Association* **81**, 446–451.

Zellner, A., and Chetty, V. K. (1965). Prediction and decision problems in regression models from the Bayesian point of view. *Journal of the American Statistical Association* **60**, 608–616.

CHAPTER 8

Screening tests for detecting a characteristic

Often it is necessary to decide whether a characteristic exists in a subject or item based on some screening test. We shall consider that the test is binary in that it indicates T (or positive) for the presence of the characteristic, $\overline{T}$ (or negative) for its absence. The merit of the test will depend on its accuracy in determining whether or not the characteristic exists in items or individuals to whom the test is administered that either possess the characteristic or do not.

The probability that an individual who has tested positive has the characteristic is termed the predictive value positive of the test. This will depend on the prevalence of the characteristic in the specified population from which an item is drawn at random. It is similar for the predictive value negative. In the medical area, screening or diagnostic tests are often used to attempt to confirm or indicate the preclinical or presymptomatic presence or near future onset of a disease (e.g., Gastwirth, 1987). Uses also abound in technology and a variety of other areas. We confine ourselves to a simple binary diagnostic test and develop the necessary tools for the full analysis of this situation. Then we shall consider two binary diagnostic tests given simultaneously or in sequence. Here the framework for the estimation of the attributes of the tests and their optimal administration for screening programs is considered. Costs of the tests and the losses associated with the various kinds of correct and incorrect decisions are also taken into account. In all that follows we assume that a screening test is always capable of assigning every individual of a particular population as a positive or negative for the characteristic, even though this may not always be the case with all tests in actual use.

8.1   Binary screening test

We first define the following probabilities:

$\Pr(C) = \pi$, the probability that a randomly drawn individual from the population exhibits characteristic C is called the prevalence;

$\Pr(T|C) = \eta$, the probability that the test correctly asserts the presence of C, is called the sensitivity;

$\Pr(\overline{T}|\overline{C}) = \theta$, the probability that the test correctly asserts the absence of $\overline{C}$ is called the specificity;

$\Pr(C|T) = \pi\eta/\{\pi\eta + (1 - \pi)(1 - \theta)\} = \psi$, the probability that the characteristic is present given the test indicates its presence is called the predictive value positive (PVP);

$\Pr(\overline{C}|\overline{T}) = (1 - \pi)\theta/\{\pi(1 - \eta) + (1 - \pi)\theta\} = \overline{\psi}$, the probability that the characteristic is absent given the test indicates its absence is called the predictive value negative (PVN).

Given the first three values π, η, and θ the last two are obtained as a simple application of Bayes theorem.

8.2   Predictive values positive and negative when the parameters are unknown

Precise knowledge of π, η, and θ is often unavailable, or only partially so, in terms of samples from various subpopulations within the relevant population. Normally θ and η will be unknown and their estimation would require independent random samples from two reference populations; the test is applied to n individuals or units known to have the characteristic, and also to $\bar{n}$ individuals known to be free of the characteristic. Assuming that r out of n yield T in the first sample and $\bar{r}$ out of $\bar{n}$ yield $\overline{T}$ in the second, we obtain the likelihood

$$L(\eta, \theta) \propto \eta^r (1 - \eta)^{n-r} \theta^{\bar{r}} (1 - \theta)^{\bar{n}-\bar{r}}.$$

If π is unknown, and an independent sample of size υ from the entire population is available, we have the additional likelihood contribution

$$L(\pi) \propto \pi^{t_c}(1 - \pi)^{t - t_c}, \tag{8.1}$$

where t_c represents the number of individuals having C. If this is unavailable, another independent sample of size s from the population could be obtained. In this sample we ascertain the number t that test T. Since

$$\Pr(T) = \pi\eta + (1 - \pi)(1 - \theta),$$

we obtain the joint likelihood

$$L(\eta, \theta, \pi) \propto \eta^r(1 - \eta)^{n - r}\theta^{\bar{r}}(1 - \theta)^{\bar{n} - \bar{r}}\{\pi\eta + (1 - \pi)(1 - \theta)\}^t$$
$$\times \{\pi(1 - \eta) + (1 - \pi)\theta\}^{s - t}.$$

Given a prior density, $p(\eta, \theta, \pi)$, and with $d = (r, n, \bar{r}, \bar{n}, t_c, v)$, the joint posterior density is obtained in the first case as

$$p(\eta, \theta, \pi | d) \propto p(\eta, \theta, \pi)L(\theta, \eta)L(\pi).$$

With $d = (r, n, \bar{r}, \bar{n}, t, s)$, we obtain for the second case

$$p(\eta, \theta, \pi | d) \propto p(\eta, \theta, \pi)L(\theta, \eta, \pi).$$

In both cases, the main quantities of interest are the predictive probabilities

$$\Pr(C | T, d) = E(\psi | T, d) = E(\pi\eta | d)$$
$$\times \{E(\pi\eta | d) + E((1 - \pi)(1 - \theta) | d)\}^{-1}, \tag{8.2}$$

and its similar counterpart

$$\Pr(\bar{C} | \bar{T}, d) = E(\bar{\psi} | \bar{T}, d)$$
$$= E(\theta(1 - \pi) | d)$$
$$\times \{E(\theta(1 - \pi) | d) + E(\pi(1 - \eta) | d)\}^{-1}, \tag{8.3}$$

both obtained by simple application of Bayes Theorem.

Clearly, the predictive probability is the same for all exchangeable subjects. Hence, when a positive result is obtained, a decision regarding various alternative actions depends on the particular subject's costs or losses. Assume that there are a actions and that for action i, the losses for a particular individual are l_{iC} and $l_{i\bar{C}}$, the loss associated with taking action i when C is true and of taking action i when $\bar{C}$ is true, respectively; $i = 1, 2, \ldots, a$. Then one need only minimize

$$E(L_i|T) = (l_{iC} - l_{i\bar{C}})\Pr(C|T, d) + l_{i\bar{C}}$$

with respect to i, to obtain the optimal action. The predictive probability of C is the same for any individual that tests positive, but individuals may very well differ in the utilities (or costs) they would assign. If costs for each action, say c_i, are in the same units as the losses, then we would minimize

$$E(L_i|T) + c_i.$$

A similar analysis holds for those that tested negative. Of course there may be individuals who would take the same action irrespective of the test result—these individuals can save the expense of taking the test. Since the posterior densities of π, η, θ can be quite complicated, approximations have been developed by Johnson and Gastwirth (1991).

8.3   Prediction for a future value

There are four other calculations of some interest (Geisser, 1987). First, consider the situation where in a screening program a hospital administrator receives only individuals who are classified as T, that is the laboratory does not inform him how many were $\bar{T}$. These individuals are assumed to be independently sampled from the same population as those that were in the training sample. Presumably, these, say Z_T, individuals will exhibit either C or $\bar{C}$ within some fixed time. In order to make appropriate preparations, the administrator would like to ascertain the number, Z_{TC} of those Z_T, who will need treatment for C. In such situations, where the number of those who are $\overline{T}$ is censored,

$$\Pr(Z_{TC} = m | Z_T = k, d) = \binom{k}{m} \int \psi^m (1 - \psi)^{k-m} \, dP(\eta, \theta, \pi | d)$$

$$= \binom{k}{m} \int \psi^m (1 - \psi)^{k-m} \, dP(\psi | d). \quad (8.4)$$

For sufficiently large Z_T

$$\Pr\left(\frac{Z_{TC}}{Z_T} \leq y | d \right) \doteq \Pr[\psi \leq y | d],$$

since by de Finetti's theorem as Z_T grows

$$\lim \frac{Z_{TC}}{Z_T} = \psi$$

with distribution function $P(\psi | d)$. If, on the other hand, a total sample of J individuals were screened and Z_T of them were T, we obtain

$$\Pr(Z_{TC} = m | Z_T = k, J, d) = \binom{k}{m} \int \psi^m (1 - \psi)^{k-m}$$

$$\times dP(\psi | Z_T = k, J, d), \quad (8.5)$$

which differs from (8.4) because of the additional information. There may also be interest in the predictive probability for Z_T, the number that test T out of J,

$$\Pr(Z_T = k | J, d) = \binom{J}{k} \int \{\pi \eta + (1 - \pi)(1 - \theta)\}^k$$

$$\times \{\pi(1 - \eta) + (1 - \pi)\theta\}^{J-k} \, dP(\pi, \eta, \theta | d).$$

Finally, the probability for the number Z_C that have the condition, among the J individuals that are to be screened, is calculated as

$$\Pr(Z_C = l | J, d) = \binom{J}{l} \int \pi^l (1 - \pi)^{J-l} \, dP(\pi | d).$$

All of these calculations require a prior distribution for θ, η, and π and numerical evaluation of the requisite integrals by one means or another. With regard to a decision framework, we could examine the problem from the point of view of the hospital administrator or, more generally, society. We may wish to assume that all individuals are exchangeable and are to be provided for equally, and thus that there is a single set of societal losses. Suppose l_{im} is the loss associated with assuming that i individuals require treatment when in fact m will. Then

$$E(L_i) = \sum_{m=0}^{Z_T} \Pr(Z_{TC} = m \mid Z_T, d) l_{im}.$$

Hence we would find the i^* that minimizes the above expectation. Let c be the actual economic cost for making treatment available to an individual, b be the benefit of treating someone who has the characteristic, d the extra cost associated with preparing for unnecessary treatment, and finally e the cost of not having treatment available for someone with the condition. Then a possibly useful set of values for l_{im} is

$$l_{im} = \begin{cases} ic - mb + (i-m)d & \text{for } i > m \\ mc - mb & \text{for } i = m \\ ic - ib + (m-i)e & \text{for } i < m. \end{cases}$$

The simplest case resulting from this is a comparison of the losses of treating no one to treating everyone. We calculate

$$E(L_0) = eE(Z_{TC})$$

$$E(L_{Z_T}) = (c + d)Z_T - (b + d)E(Z_{TC}).$$

Note as $E(Z_{TC}) \to 0$, $E(L_0) \to 0$ indicating that no preparation is dominant while if $E(Z_{TC}) \to Z_T$ then complete preparation will dominate if and only if $c < b + e$. Calculations for the L_i, $0 < i < Z_T$ are more complicated and could require the aid of a computer.

8.4 Comparison of tests

Before we compare tests we would like to have an indication of when a test is diagnostically useful. Clearly unless a given test has the minimal property that $\psi = \Pr(C|T) > 0.5$ and $\bar{\psi} = \Pr(\bar{C}|\bar{T}) > 0.5$, it is useless. One would of course prefer these probabilities to be considerably greater for the test to be diagnostically useful, say $\psi > \gamma > 0.5$ and $\bar{\psi} > \bar{\gamma} > 0.5$, for some fixed values γ and $\bar{\gamma}$. In particular when using the test for a given individual, when η, θ, and π are unknown, we would prefer that the predictive probabilities

$$\Pr(C|T, d) = E(\psi|T, d) > \gamma > 0.5$$
$$\Pr(\bar{C}|\bar{T}, d) = E(\bar{\psi}|\bar{T}, d) > \bar{\gamma} > 0.5.$$

For a large number of future individuals, appropriate criteria for a useful diagnostic test would be

$$\Pr[\psi > \gamma] > \delta > 0.5$$
$$\Pr[\bar{\psi} > \bar{\gamma}] > \bar{\delta} > 0.5$$

or jointly

$$\Pr[\psi > \gamma, \bar{\psi} > \bar{\gamma}] > \delta_0 > 0.5.$$

In deciding which test (if only one test can be used) of the two T_1 or T_2 to administer, clearly we would prefer T_1 to T_2 if

$$\psi_1 > \psi_2 \quad \text{and} \quad \bar{\psi}_1 > \bar{\psi}_2.$$

When ψ_i and $\bar{\psi}_i$, $i = 1, 2$, are unknown, the predictive probabilities

$$E(\psi_1|d) > E(\psi_2|d)$$
$$E(\bar{\psi}_1|d) > E(\bar{\psi}_2|d)$$

are sufficient for T_1 to be preferred to T_2. These expectations are the predictive probabilities regarding a new individual who has tested positive for the first inequality or negative for the second inequality. In dealing with a large number of future individuals one could use as a comparison of the two tests

$$\Pr\left[\psi_1 > \psi_2, \bar{\psi}_1 > \bar{\psi}_2\right] > 0.5$$

to indicate the superiority of T_1 to T_2.

8.5 Two tests

In this section, we allow for the availability of two distinct tests that may be administered either simultaneously or sequentially in a mass screening program. There are a number of different decision rules one can follow regarding how to proceed. If one were to administer two tests simultaneously, one could assert that the characteristic exists or preexists if both tests are positive or if either test is positive. A decision could be made to ignore one of the test results and to simply designate C or $\bar{C}$ according to a single test's result. One could administer the tests sequentially and allow for the possibility of administering one test first and then either stopping if the result were positive, or proceeding to a second test if the result were negative. The final decision would be to designate as C either a positive on the first test or a negative on the first and a positive on the second. Another sequential possibility is to give a single test and to designate $\bar{C}$ if the result is negative, otherwise give the second test and designate C only if both tests are positive.

We denote our single test results as $(T_i, \bar{T}_i)$ $i = 1, 2$, to denote positive or negative, and introduce notation to indicate the above decision rules. There are eight possibilities that we will discuss. They are listed in Table 8.1. Decision rule, R_8 say, resulted in a positive result, T, if Test 2 was administered first and was positive or if it was negative and then Test 1 was subsequently positive. The omitted possibilities correspond to rules that can never be optimal for a sensible loss structure; for example, assert C if Test 1 is negative, denoted as $\bar{T}$ (Geisser and Johnson, 1992).

There are expenses associated with the tests, and losses in making wrong decisions. We defer discussion of expenses for the time being. Here, we define a function that reflects the various

Table 8.1 *Decision Rules*

Rule	Decision rule (assert C if)	Notation
R_1	Test 1 is positive	T_1
R_2	Test 2 is positive	T_2
R_3	Both tests positive (simultaneous tests)	$(T_1 T_2)$
R_4	Either test positive (simultaneous tests)	$(T_1 \cup T_2)$
R_5	Both tests positive (sequential tests)	$T_1 T_2$
R_6	Both tests positive (sequential tests)	$T_2 T_1$
R_7	Either test positive (sequential tests)	$T_1 \cup \bar{T}_1 T_2$
R_8	Either test positive (sequential tests)	$T_2 \cup \bar{T}_2 T_1$

losses associated with correct and incorrect decisions. Later we will define a function that reflects the actual costs or expense of administration connected with the decision rules.

We define our loss function in Table 8.2. For example, the cost of a positive decision when C is present is l_{11}. The first subscript of l_{ij} denotes the decision, T or $\bar{T}$, and the second denotes presence or absence of the characteristic.

Table 8.2 *Loss Function*

		True state	
		C	$\bar{C}$
Decision rule	T	l_{11}	l_{10}
Outcome	$\bar{T}$	l_{01}	l_{00}

When C denotes some serious condition, there are two points of view that one can take regarding this cost function; one from the perspective of society and the other from that of the individual. From a societal standpoint, it could be sensible to argue that

$$l_{11} \le l_{00} < l_{10} \le l_{01}.$$

Subsequently, we consider a situation where the cost associated with infecting an individual by transfusion of contaminated blood (containing a communicable virus) is assumed to be high, say $l_{01} = \$10^6$. While there are costs associated with false positive results, they are expected to be much less than $\$10^6$. The individual involved would generally not be informed of a positive result unless it was confirmed by additional testing to be positive, thus the loss l_{10} would be expected to range from a minimum amount (the cost of the initial test) to further costs, depending on how many confirmatory tests were equivocal before a definite decision could be made regarding the infection status. In the event that individuals were informed as to the results of the screening test, these costs could be much higher. The ordering could vary from one individual to another. Drug testing is a good example of a situation where the ordering would vary depending on one's perspective. In general, it would only be reasonable to assume $l_{ii} < l_{kj}$ for all $k \neq j$.

A generalization of our notation from Section 8.2 regarding the conditional probabilities for the various outcomes of the two tests is introduced in Table 8.3. Thus $\Pr(T_1, T_2 | C) = \eta_{11}$, $\Pr(\bar{T}_1, T_2 | \bar{C}) = \theta_{01}$, etc. The first subscript corresponds to a positive or negative result on Test 1 and the second denotes the same information for Test 2. We define $\Pr(C) = \pi$ as before, and the predictive values positive and negative for decision rule R_i as

$$\Pr(C|T) = \psi_i \quad \text{and} \quad \Pr(\bar{C}|\bar{T}) = \bar{\psi}_i, \quad \text{for } i = 1, 2, \ldots, 8.$$

Table 8.3 *Conditional Probabilities for the Outcomes of the Two Tests*

	C		$\bar{C}$	
	T_2	$\bar{T}_2$	T_2	$\bar{T}_2$
T_1	η_{11}	η_{10}	θ_{11}	θ_{10}
$\bar{T}_1$	η_{01}	η_{00}	θ_{01}	θ_{00}

For each of the eight joint tests in Table 8.1, define the sensitivity and specificity as

$$\eta_i = \Pr(T|C), \qquad \theta_i = \Pr(\bar{T}|\bar{C}) \qquad i = 1, \ldots, 8.$$

For decision rule R_i, the expected loss is

$$E(L|R_i) = \pi \eta_i [l_{11} - l_{01}] + (1 - \pi)\theta_i [l_{00} - l_{10}] + \pi l_{01}$$
$$+ (1 - \pi)l_{10}. \tag{8.6}$$

Table 8.4 $\Pr(T|C)$ and $\Pr(\bar{T}|\bar{C})$ for the First Four Decision Rules

Rule	$\Pr(T|C)$ (sensitivity)	$\Pr(\bar{T}|\bar{C})$ (specificity)
R_1	$\eta_1 = \eta_{11} + \eta_{10}$	$\theta_1 = \theta_{00} + \theta_{01}$
R_2	$\eta_2 = \eta_{11} + \eta_{01}$	$\theta_2 = \theta_{00} + \theta_{10}$
R_3	$\eta_3 = \eta_{11}$	$\theta_3 = \theta_{00} + \theta_{01} + \theta_{10}$
R_4	$\eta_4 = \eta_{11} + \eta_{10} + \eta_{01}$	$\theta_4 = \theta_{00}$

The values of $\Pr(T|C)$ and $\Pr(\bar{T}|\bar{C})$ for each of the first four decision rules are presented in Table 8.4. Note that, irrespective of prevalence and losses resulting from either false positive or negative results, if $\eta_{10} = \eta_{01} = 0$, $(T_1 T_2)$ is optimal. If $\theta_{01} = \theta_{10} = 0$, $(T_1 \cup T_2)$ is optimal. If $\eta_{10} = \theta_{01} = 0$, T_2 is optimal, and if $\eta_{01} = \theta_{10} = 0$, T_1 is optimal. Finally, it follows easily that the values for R_5 and R_6 are equivalent to those for R_3, and similarly R_7 and R_8 are equivalent to R_4. Consideration of other rules is precluded since it is sufficient that other rules cannot be optimal if the following sensible conditions hold, namely, $\max(l_{11}, l_{00}) < \min(l_{10}, l_{01})$, $\eta_{11} > \eta_{ij} > \eta_{00}$ and $\theta_{00} > \theta_{ij} > \theta_{11}$ for $i \neq j = 0$. When the parameters are unknown we assume that their joint posterior distribution is sufficiently concentrated on the parameter space given above that similarly the other rules remain inadmissible.

For a mass screening program, such as contemplated for certain diseases, it would be desirable to use the optimal rule. Now rule R will be preferred to R^* if the expected loss under R is less than that under R^*. Denote this as $R > R^*$. It is straightforward to show that if $k = (l_{10} - l_{00})/(l_{10} - l_{00} + l_{01} - l_{11})$, then, in the more easily recognizable notation, e.g, R_3 is $(T_1 T_2)$,

$$T_2 > (T_1 \cup T_2) \Leftrightarrow (T_1 T_2) > T_1 \Leftrightarrow \Pr(C|T_1\bar{T}_2) < k$$

$$T_1 > (T_1 \cup T_2) \Leftrightarrow (T_1 T_2) > T_2 \Leftrightarrow \Pr(C|\bar{T}_1 T_2) < k$$

$$(T_1 T_2) > (T_1 \cup T_2) \Leftrightarrow \Pr(C|\bar{T}_1 T_2 \cup T_1\bar{T}_2) > k$$

$$T_2 > T_1 \Leftrightarrow \frac{\Pr(C, T_1\bar{T}_2) - \Pr(C, \bar{T}_1 T_2)}{\Pr(T_1\bar{T}_2) - \Pr(\bar{T}_1 T_2)} < k$$

$$\Pr(C|T_1\bar{T}_2) < \Pr(C|T_1\bar{T}_2 \cup \bar{T}_1 T_2)$$
$$< \Pr(C|\bar{T}_1 T_2) \Leftrightarrow \Pr(C|T_1\bar{T}_2)$$
$$< \Pr(C|\bar{T}_1 T_2). \tag{8.7}$$

Note that $k = 1/2$ when $l_{10} - l_{00} = l_{01} - l_{11}$ or more particularly if $l_{00} = l_{11}$ and $l_{10} = l_{01}$. Suppose that $\Pr(C|T_1\overline{T}_2) < k < \Pr(C|\overline{T}_1T_2)$. Then it follows that

$$T_2 > (T_1 \cup T_2) > T_1; \quad \text{and} \quad T_2 > (T_1T_2) > T_1,$$

thus T_2 would be the optimal procedure if the actual monetary cost of testing were irrelevant. In fact, it is also the case that if T_2 is optimal, then the above condition must hold, due to (8.7), so the condition is necessary and sufficient. The following necessary and sufficient conditions are shown using the previous results:

$$T_1 \text{ is optimal} \Leftrightarrow \Pr(C|\overline{T}_1T_2) < k < \Pr(C|T_1\overline{T}_2)$$

$$T_2 \text{ is optimal} \Leftrightarrow \Pr(C|T_1\overline{T}_2) < k < \Pr(C|\overline{T}_1T_2)$$

$$(T_1T_2) \text{ is optimal} \Leftrightarrow \Pr(C|T_1\overline{T}_2) < k \quad \text{and} \quad \Pr(C|\overline{T}_1T_2) < k$$

$$(T_1 \cup T_2) \text{ is optimal} \Leftrightarrow \Pr(C|T_1\overline{T}_2) > k \quad \text{and} \quad \Pr(C|\overline{T}_1T_2) > k.$$

$$(8.8)$$

These conditions translate easily into equivalent statements regarding the parameters $\{\eta_{ij}, \theta_{ij}\}$ and π.

With the values in Table 8.3 fixed and known, it is possible to partition the space of (k, π) values into regions for which the various rules are optimal. Define $k^* = k/(1-k)$, $a = \theta_{01}/\eta_{01}$, $b = \theta_{10}/\eta_{10}$, $a^* = \theta_{01}/(\theta_{01} + \eta_{01})$, $b^* = \theta_{10}/(\theta_{10} + \eta_{10})$. Then

$$T_1 \text{ is optimal} \Leftrightarrow \pi < \frac{ak^*}{1 + ak^*}, \qquad \pi > \frac{bk^*}{1 + bk^*}$$

$$T_2 \text{ is optimal} \Leftrightarrow \pi < \frac{bk^*}{1 + bk^*}, \qquad \pi > \frac{ak^*}{1 + ak^*}$$

$$(T_1T_2) \text{ is optimal} \Leftrightarrow \pi < \min\left(\frac{ak^*}{1 + ak^*}, \frac{bk^*}{1 + bk^*}\right)$$

$$(T_1 \cup T_2) \text{ is optimal} \Leftrightarrow \pi > \max\left(\frac{ak^*}{1 + ak^*}, \frac{bk^*}{1 + bk^*}\right).$$

$$(8.9)$$

If $a < b$, the regions of optimality are given in Figure 8.1. If $a > b$, then the middle region corresponds to values of (k^*, π) for which T_1 is optimal; T_2 would thus never be optimal when $a < b$.

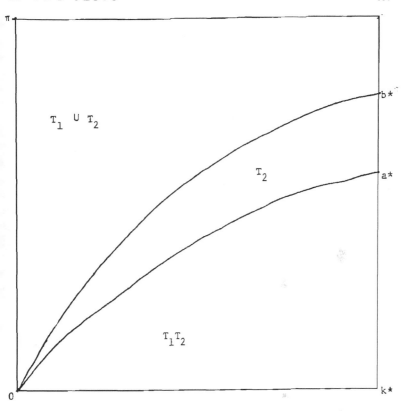

Figure 8.1   *Regions of optimality*

Now assume the parameters are unknown and that prior information is available for $(\{\eta_{ij}\}, \{\theta_{ij}\}, \pi)$ in the form of independent Dirichlet distributions, namely

$$p(\{\eta_{ij}\}) \propto \prod_{i,j} \eta_{ij}^{\alpha_{ij}-1}; \qquad \eta_{ij} \ge 0, \qquad \sum_{i,j} \eta_{ij} = 1, \qquad \sum_{i,j} \alpha_{ij} = \alpha,$$

$$p(\{\theta_{ij}\}) \propto \prod_{i,j} \theta_{ij}^{\bar{\alpha}_{ij}-1}; \qquad \theta_{ij} \ge 0, \qquad \sum_{i,j} \theta_{ij} = 1, \qquad \sum_{i,j} \bar{\alpha}_{ij} = \bar{\alpha},$$

$$p(\pi) \propto \pi^{\gamma-1}(1-\pi)^{\bar{\gamma}-1}, \qquad 0 \le \pi \le 1.$$

Assume that data of two types are available. The first type consists of "training data," where individuals known to have C and individuals known to be $\bar{C}$ are tested and a decision made as to T or $\bar{T}$ in each case. The likelihood corresponding to such data is

$$L\big(\{\eta_{ij}\},\{\theta_{ij}\}\big) = \prod_{i,j}\eta_{ij}^{r_{ij}}\theta_{ij}^{\bar{r}_{ij}}; \qquad \sum_{i,j}r_{ij} = n, \qquad \sum_{i,j}\bar{r}_{ij} = \bar{n}.$$

The second type of data contains information for π; such data can either be direct or indirect, as in the one test case discussed in Section 8.2. Direct data for π are assumed to arise from a binomial sample for π; the corresponding likelihood is the same as in (8.1). Indirect data for π are obtained from screening data on individuals whose status with respect to C is unknown. This type of data is especially appropriate for small π, since direct estimates of π in such cases would often be prohibitively expensive and/or impossible to obtain. We assume indirect data for π is in the form of a screening test from the general population under scrutiny. Many types of independent screening data are possible; for example

S_1: s_1 individuals are randomly selected and tested with Test 1 and s_2 individuals are randomly selected and tested with Test 2;

S_2: s individuals are randomly selected and tested with both Tests 1 and 2.

The likelihoods for S_1 and S_2 are

$$L_1\big(\{\eta_{ij}\},\{\theta_{ij}\},\pi\big) = \prod_{i=1}^{2}\{\pi\eta_i + (1-\pi)(1-\theta_i)\}^{t_i}$$

$$\times\{\pi(1-\eta_i) + (1-\pi)\theta_i\}^{s_i - t_i}$$

$$L_2\big(\{\eta_{ij}\},\{\theta_{ij}\},\pi\big) = \prod_{i,j}\{\pi\eta_{ij} + (1-\pi)\theta_{ij}\}^{t_{ij}}.$$

Of course, screening data could be obtained from any of the sampling schemes that correspond to the other decision rules in Table 8.1.

For the above sampling designs, the total likelihood is either

$$L(\pi)L(\{\eta_{ij}\},\{\theta_{ij}\})$$

or

$$L_i(\{\eta_{ij}\},\{\theta_{ij}\},\pi)L(\{\eta_{ij}\},\{\theta_{ij}\}) \qquad i = 1,2.$$

Under the first sampling scheme, the posteriors for $\{\eta_{ij}\}, \{\theta_{ij}\}$, and π are independent and Dirichlet, i.e., $X_{ij} \sim D(\{\theta_{ij}\})$ indicates that

$$f(\{x_{ij}\}) \propto \prod_i \prod_j x_{ij}^{\theta_{ij}}.$$

Writing d for data,

$$\{\eta_{ij}\}\,|d \sim D(\{r_{ij} + \alpha_{ij}\}), \{\theta_{ij}\}\,|d \sim D(\{\bar{r}_{ij} + \bar{\alpha}_{ij}\})$$
$$\pi|d \sim \text{Beta}(\gamma + t_C, \bar{\gamma} + v - t_C). \tag{8.10}$$

Posterior means and variances and the sensitivities and specificities of all the decision rules are easily obtained under this setup. For example, the expected sensitivity for R_1 is

$$E(\eta_1|d) = (r_{11} + r_{10} + \alpha_{11} + \alpha_{10})/(n + \alpha)$$

and the expected prevalence is

$$E(\pi|d) = (\gamma + t_c)/\bar{\gamma} + v).$$

Joint posteriors under screening data S_1 and S_2 are easily obtained, but are computationally burdensome. Simplifications in the high accuracy and low prevalence case are possible. We shall assume here that direct information is available for π, and thus the posterior distribution of all parameters will be given by (8.6). In these cases where current data are unavailable, one's prior knowledge of π in conjunction with data for the η_{ij}s and θ_{ij}s may be sufficiently precise to warrant an analysis without current data for π.

The predictive probability of C for an individual that has just tested positive according to decision rule R_i is calculated as

$$\Pr(C|T, d) = E(\psi_i|\overline{T}, d).$$

On application of Bayes Theorem, this can also be expressed as in (8.2) with (η_i, θ_i) substituted for (η, θ). Similarly, $\Pr(\overline{C}|\overline{T}, d) = E(\overline{\psi}_i|\overline{T}, d)$, the probability that an individual who tested negative on the test is actually $\overline{C}$ can be calculated as in (8.3). To obtain values from a particular rule, we simply substitute the appropriate sensitivity and specificity from Table 8.4 into (8.2) and (8.3). Under our sampling scheme for π, all expectations of products result in products of expectations due to independence. For example, with rule R_i,

$$\Pr(C|T, d) = \frac{E(\pi|d)E(\eta_i|d)}{E(\pi|d)E(\eta_i|d) + E(1 - \pi|d)E(1 - \theta_i|d)}.$$

The optimal rule in the case of unknown parameters can now be given. From (8.1) we obtain the expected loss, given rule R_i, as

$$E(L|R_i, d) = E(\pi\eta_i|d)(l_{11} - l_{01}) + E((1 - \pi)\theta_i|d)(l_{00} - l_{10})$$
$$+ E(\pi|d)l_{01} + E(1 - \pi|d)l_{10}. \qquad (8.11)$$

On substitution of particular values for the sensitivity and specificity for different rules, it follows from (8.11) that (8.7) and (8.8) must hold with all conditional probabilities replaced with corresponding predictive probabilities. For example, replace $\Pr(C|T_1\overline{T}_2)$ with $\Pr(C|T_1\overline{T}_2, d) = E(\pi\eta_{10}|d)/E(\pi\eta_{10} + (1 - \pi)\theta_{10}|d)$, etc.

8.6   Costs of administration

In any large scale screening program costs of administering the tests will be of considerable concern. For example, with regard to two diagnostic tests for the same characteristic, one may be an order of magnitude less expensive than the other. Also, there may be a differential in the cost of a simultaneous administration of both tests in contrast to their sequential administration. Among other things this may result from having to store the sample until the result from the first test is obtained, or from asking a testee to return for testing.

Once all of the actual testing costs are carefully ascertained, their incorporation into a complete decision analysis can be made without much difficulty. The major problems are in assessing in some reasonable way the original losses on a comparable monetary scale with the actual expense of testing. Another decision analytic consideration would be for a situation where a program was limited to a fixed amount of funding because of competing concerns. In this case, procedures that optimize certain well-defined benefits would be of critical interest.

Let K_i be the cost of administering Test i alone, and let K_{ij} be the cost of administering Test i followed by administering Test j. Let $K_{(12)}$ be the cost of administering both tests simultaneously. Clearly, it is reasonable to assume that $K_{ij} \geq K_{(12)} \geq \max(K_1, K_2)$. In testing for a disease, for example, there may be storage and retrieval costs for sequential sampling.

The expected cost for the eight rules are given in Table 8.5. The probabilities are conditional on the parameters. If they were unknown, then they would be conditional on d.

Table 8.5   *Expected Cost for the Eight Rules*

Rule	E(Cost)
R_1	K_1
R_2	K_2
R_3	$K_{(12)}$
R_4	$K_{(12)}$
R_5	$K_1 + (K_{12} - K_1)Pr(T_1)$
R_6	$K_2 + (K_{21} - K_2)Pr(T_2)$
R_7	$K_1 + (K_{12} - K_1)Pr(\bar{T}_1)$
R_8	$K_2 + (K_{21} - K_2)Pr(\bar{T}_2)$

Finally, define the total expected loss for decision rule R_i to be the sum of the appropriate expected loss from (8.3) or (8.11) and the appropriate expected cost from Table 8.5. We find that the total expected loss for R_5 is greater than that for R_3 if and only if $E(\text{Cost})$ for R_5 is greater than that for R_3. The same holds true for R_6 compared with R_3. R_7 is preferable to R_4 if $E(\text{Cost})$ for R_7 is less than $E(\text{Cost})$ for R_4, and the same holds true for R_8 compared with R_4. The decision as to whether to consider sequential tests versus simultaneous tests is based purely on costs. Thus if $K_{12} = K_{21} = K_{(12)}$ or more particularly if $K_{12} = K_{(12)} = K_1 + K_2$, then R_5 and R_6 will be preferable to R_3 and R_7 and R_8 will be preferable to R_4.

8.7 Numerical illustrations of screening

We consider two data sets abstracted from studies that were designed to compare different tests for detecting antibodies to the AIDS virus (Burkhardt et al., 1987; Sandstrom et al., 1985). Neither study contained information for π. In both of these studies, we utilize data for the accuracies in conjunction with independent information for π from the population of blood donors in Canada (Nusbacher et al., 1986).

For both of the AIDS studies, individual serum specimens were tested with two commercial preparations. For our purposes, weak positives and negatives were considered positive. In both studies, specimens that were weakly positive or negative on either test, or that resulted in discrepant results ($T_1\overline{T}_2$) or ($\overline{T}_1T_2$) were retested in a variety of ways, including the Western Blot test, an ultimate determination as to the true status of the specimen was ascertained. In one of the studies, a subsample of (T_1T_2) and ($\overline{T}_1\overline{T}_2$) samples was given confirmatory tests and one discrepancy was noted. From these studies, we have abstracted $2 \times 2 \times 2$ tables of counts in the form of Table 8.3. It will be assumed that these are the same tables that would have arisen if all specimens were given confirmatory tests. Since there are possible errors in the abstraction of the data, the analyses are only illustrative.

Assuming a generic measurement scale for the moment, let $l_{00} = l_{11} = 0$ and $l_{01} = 1 \geq l_{10} = q > 0$ for this study. This reflects a societal attitude that it is worse to infuse contaminated blood than it is to falsely identify a healthy donor in the AIDS examples. With these assumptions, $k = q/(1 + q)$ and $k^* = q$.

Finally, regarding the priors, we assume $\alpha_{11} = 3.9 = \bar{\alpha}_{00}$, $\alpha_{10} = \alpha_{01} = 0.5 = \bar{\alpha}_{10} = \bar{\alpha}_{01}$, and $\alpha_{00} = 0.1 = \bar{\alpha}_{11}$. Thus, our prior for $\{\eta_{ij}\}$ has a weight, $\alpha_{11} + \alpha_{10} + \alpha_{01} + \alpha_{00} = 5$, which is equivalent to a sample of size $n = 5$ for the data. The same statement holds for $\{\theta_{ij}\}$. Furthermore, our prior reflects the belief that simultaneous correct screening results are quite likely (prior probability = $3.9/5$ = 0.78) and that incorrect results are considerably less likely, but still plausible.

For information on π we utilize Canadian data obtained by Nusbacher et al. (1986). They administered a questionnaire that was designed to encourage "high risk" individuals to donate their blood anonymously for research purposes rather than for possible transfusion. Among 94,496 individuals whose blood was not utilized for research purposes, 405 samples tested positive on an ELISA screening test, and among those, 14 were confirmed positive by Western Blot. If we assume the remaining 94,091 negative samples were actually uncontaminated, then these data could be treated as a Bin $(94496, \pi)$ sample. The prevalence π is the proportion of contaminated samples in the Canadian blood that is available for transfusion. With an improper $B(1,0)$ prior for π, $E(\pi|d) = 0.000148$. Alternatively, one could simply assume a beta prior with mean 0.000148 and an appropriate standard deviation.

Now Burkhardt et al. (1987) sampled 503 individuals and tested each serum specimen with a drug company "A" ELISA test and a drug company "D" ELISA test. The data on accuracies in Table 8.6 were abstracted from their study.

Table 8.6 *Data on Accuracies*

| | A-ELISA | | | |
| | C | | $\bar{C}$ | |
	T	$\bar{T}$	T	$\bar{T}$
D-ELISA				
T	92	0	8	9
$\bar{T}$	1	0	23	370

We do not assume specific information for π, nor a specific value for q as yet, but we do assume that a prior for π is available and direct information for π could be obtained in the form (8.1). The resulting posterior for π is beta. Utilizing the data displayed in Table 8.6, we obtain independent Dirichlet posteriors for $(\{\eta_{ij}\}\{\theta_{ij}\}, \pi)$.

The conditional predictive probabilities of positive and negative results for rules R_1-R_4 are listed in Table 8.7 where the notation is $D = T_1$, $A = T_2$. These probabilities are calculated as $\Pr(T|C, d) = E(\eta|d)$, $\Pr(\bar{T}|\bar{C}, d) = E(\theta|d)$.

Table 8.7 *Conditional Predictive Probabilities of Positive and Negative Results for R_1-R_4*

Rule	Assert C	$\Pr(T|C, d)$ (sensitivity)	$\Pr(\bar{T}|\bar{C}, d)$ (specificity)
R_1	D is T	0.984	0.958
R_2	A is T	0.994	0.924
R_3	(DA) both T	0.979	0.981
R_4	$(D \cup A)$ either T	0.999	0.901

Considering the optimality of rules R_1-R_4, we obtain

$$a = E(\theta_{01}|d)/E(\eta_{01}|d) = 3.70$$

$$b = E(\theta_{10}|d)/E(\eta_{10}|d) = 4.50.$$

Optimality is determined according to (8.6) and is viewed pictorially by consideration of Figure 8.1 with $a^* = 0.79$, $b^* = 0.82$, $k^* = q$, and $E(\pi|d)$ substituted for π. The D-ELISA alone would not be optimal for any q regardless of the value of $E(\pi|d)$.

If we assume the posterior for π has mean 0.000148, as discussed above for the Canadian transfusion pool, it remains to consider a before making a decision regarding the choice among rules R_1-R_4, without regard to costs of administering the tests.

Suppose for the sake of illustration that the value $\$10^6$ is attached to the life of an individual transfused with contaminated blood. In this instance, we find that the rule (DA) is preferable if $q > \$40$, the rule $(D \cup A)$ is preferable if $q < \$33$, and the rule A is preferable if $\$33 < q < \$40$.

We now consider the expenses associated with the administration of these tests. For simplicity, assume $K_1 = K_2 = K$, $K_{(12)} =$

$K_{12} = K_{21} = 2K$. Then

$$E(\text{Cost}|R_1) = E(\text{Cost}|R_2) = K,$$
$$E(\text{Cost}|R_3) = E(\text{Cost}|R_4) = 2K,$$
$$E(\text{Cost}|R_5) = K\{1 + \Pr(D|d)\},$$
$$E(\text{Cost}|R_6) = K\{1 + \Pr(A|d)\},$$
$$E(\text{Cost}|R_7) = K\{1 + \Pr(\overline{D}|d)\},$$
$$E(\text{Cost}|R_8) = K\{1 + \Pr(\overline{A}|d)\},$$

where

$$\Pr(D|d) = 0.941 E(\pi|d) + 0.042,$$
$$\Pr(A|d) = 0.918 E(\pi|d) + 0.076.$$

Thus for the situation above with $E(\pi|d) = 0.000148$,

$$E(\text{Cost}|R_5) \doteq 1.042 K, \qquad E(\text{Cost}|R_6) \doteq 1.076 K$$
$$E(\text{Cost}|R_7) \doteq 1.958 K, \qquad E(\text{Cost}|R_8) \doteq 1.924 K.$$

Suppose $K = \$1$ and $q = \$25$. Without regard to the cost of administering a test, $(D \cup A)$ is optimal. The total expected losses, per individual, for the eight rules are \$4.42, \$3.79, \$5.58, \$4.62, \$4.46, \$3.87, \$5.54, and \$4.54, respectively. Thus A would be optimal overall. If on the other hand $K = \$5$ and $q = \$50$, then (DA) is optimal without considering test expense. The total expected costs for the eight rules are \$11.57, \$13.49, \$15.01, \$20.05, \$10.22, \$10.39, \$19.84, and \$19.67, respectively. Thus the sequential rule (DA) would be optimal.

We finally list the PVP and $1 - \text{PVN}$ values for the rules (Table 8.8).

Table 8.8 *PVP and* $1 - PVN$ *Values for the Rules*

Rule	$\Pr(C|T, d) = \text{PVP}$	$\Pr(C|\overline{T}, d) = 1 - \text{PVN}$
R_1	0.0035	2.4×10^{-6}
R_2	0.0019	0.8×10^{-6}
R_3	0.0076	3.1×10^{-6}
R_4	0.0015	0.2×10^{-6}

References

Burkhardt, U., Mertens, Th., and Eggers, H. J. (1987). Comparison of two commercially available anti-HIV ELISAs: Abbott HTLV III EIAc and DuPont HTLV III-ELISAs. *Journal of Medical Virology* **23**, 217–224.

Gastwirth, J. L. (1987). The statistical precision of medical screening tests. *Statistical Science* **2**, 213–238.

Geisser, S. (1987). Comments on Gastwirth, the statistical precision of medical screening tests. *Statistical Science* **2**, 231–232.

Geisser, S., and Johnson, W. (1992). Optimal administration of dual screening tests for detecting a characteristic, with special relevance to low prevelance diseases. *Biometrics* **48**, 839–852.

Johnson, W. O., and Gastwirth, J. L. (1991). Bayesian inference for medical screening tests: Approximations useful for the analysis of AIDS data. *Journal of the Royal Statistical Society B* **53**(2), 427–440.

Nusbacher, J., Chiavetta, J., Naiman, R., Buchner, B., Scalia, V., and Horst, R. (1986). Evaluation of a confidential method of excluding blood donors exposed to human immunodeficiency virus. *Transfusion* **26**, 539–541.

Sandstrom, E. G., Schooley, R. T., Ho, D. D., Byington, R., Sarnyadharan, M. G., MacLane, M. E., Essex, M., Gallo, R. C., and Hirsch, M. S. (1985). Detection of anti-HTLV-III antibodies by indirect immunofluorescence using fixed cells. *Transfusion* **25**, 308–312.

CHAPTER 9

Multivariate normal prediction

When modeling a continuous random variable the univariate normal distribution has been more frequently and successfully used in applications than any other distribution. In dealing with continuous p-dimensional vectorial random variables, the potential number of joint distributions, depending on various conditional and marginal distributions, is combinatorially increasing in the dimension p. Nevertheless, there are many data sets where all of the marginal and conditional distributions are sufficiently close to normality such that the multivariate normal distribution can be considered a reasonable model for use in problems involving prediction. In this chapter, we shall restrict our attention to this model.

9.1 Multivariate normal prediction of a future observation

Let X be a p-dimensional random variable having density function

$$f_X(x) = \frac{|\Sigma|^{-1/2}}{(2\pi)^{p/2}} e^{-(1/2)\operatorname{tr}\Sigma^{-1}(x-\mu)(x-\mu)'}$$

i.e., an $N(\mu, \Sigma)$ distribution with

$$\mu = \begin{pmatrix} \mu_1 \\ \vdots \\ \mu_p \end{pmatrix} \quad \text{and} \quad \Sigma = (\sigma_{ij}).$$

Let $x_1, x_2, \ldots, x_N$ be a random sample of p-dimensional vector observations on $N(\mu, \Sigma)$. To obtain a subjective prior for $\frac{1}{2}p(p+3)$ parameters of the multivariate normal distribution is basically a hopeless task for any $p > 2$ unless some very specific situation obtains. In the absence of such we shall use

$$p(\mu, \Sigma^{-1}) \propto |\Sigma|^{(p+1)/2}$$

first suggested by Geisser and Cornfield (1963). This can be derived in a number of ways, one of which is as a limiting case of a conjugate prior on μ and Σ^{-1}. It can also be shown to satisfy the invariance criterion of Jeffreys if it is slightly stretched to imply prior independence of μ and Σ^{-1} and the use of only

$$p(\mu, \Sigma^{-1}) \propto |\mathbf{I}(\Sigma^{-1})|^{1/2} = |\Sigma|^{(p+1)/2}$$

rather than

$$p(\mu, \Sigma^{-1}) \propto |\mathbf{I}(\mu, \Sigma^{-1})|^{1/2}$$

as might be implied by the original Jeffreys' criterion. The former provides invariance for μ alone and for Σ^{-1} alone but not jointly, while the latter provides joint invariance for μ and Σ^{-1} [see Box and Tiao (1973) for a derivation from Jeffreys' criterion]. Another justification for this prior will be given in a predictive context afterward. Now from the random sample of size N the likelihood can be written as

$$L(\mu, \Sigma^{-1}) \propto |\Sigma^{-1}|^{N/2} e^{-(1/2)\operatorname{tr}\Sigma^{-1}[(N-1)S + N(\bar{x}-\mu)(\bar{x}-\mu)']}$$

where

$$\bar{x} = \frac{1}{N} \sum_{\alpha=1}^{N} x_\alpha \quad \text{and} \quad S = \frac{1}{N-1} \sum_{\alpha=1}^{N} (x_\alpha - \bar{x})(x_\alpha - \bar{x})'.$$

Hence the posterior density depends only on $\bar{x}$ and S and is

$$p(\mu, \Sigma^{-1} | \bar{x}, S) \propto L(\mu, \sigma^{-1}) |\Sigma|^{(p+1)/2}$$

$$\propto |\Sigma^{-1}|^{(N-p-1)/2} e^{-(1/2)\operatorname{tr}\Sigma^{-1}[(N-1)S + N(\bar{x}-\mu)(\bar{x}-\mu)']}.$$

Further

$$f(x_{N+1}|\bar{x}, S)$$

$$= \int \int f(x_{N+1}|\mu, \Sigma^{-1}) p(\mu, \Sigma^{-1}|\bar{x}, S)\, d\mu\, d\Sigma^{-1}$$

$$\alpha \int \int |\Sigma^{-1}|^{(N-p)/2}$$

$$\times e^{-(1/2)\mathrm{tr}\, \Sigma^{-1}[(N-1)S + N(\bar{x}-\mu)(\bar{x}-\mu)' + (x_{N+1}-\mu)(x_{N+1}-\mu)']}\, d\mu\, d\Sigma^{-1}.$$

Now

$$N(\bar{x}-\mu)(\bar{x}-\mu)' + (x_{N+1}-\mu)(x_{N+1}-\mu)'$$

$$= (N+1)(\bar{d}-\mu)(\bar{d}-\mu)' + \frac{N}{N+1}(\bar{x}-x_{N+1})(\bar{x}-x_{N+1})'$$

where $\bar{d} = (N+1)^{-1}(N\bar{x} + x_{N+1})$. Hence it is seen that

$$f(x_{N+1}|\bar{x}, S)$$

$$= \left[\frac{N}{(N+1)\pi} \right]^{p/2} \frac{\Gamma(N/2)|(N-1)S|^{(N-1)/2}}{\Gamma[(N-p)/2]}$$

$$\times |(N-1)S + [N/(N+1)](\bar{x}-x_{N+1})(\bar{x}-x_{N+1})'|^{-N/2}.$$

$$(9.1)$$

From the above or by other methods it can be shown that $[(N-p)N]/[p(N^2-1)](X_{N+1}-\bar{x})'S^{-1}(X_{N+1}-\bar{x})$ is distributed as an $F(p, N-p)$ variate. Therefore a $1-\alpha$ predictive probability region for the as yet to be observed X_{N+1} can be found through

$$\Pr\left[(X_{N+1}-\bar{x})'S^{-1}(X_{N+1}-\bar{x})\right] \le \frac{p(N^2-1)}{N(N-p)} F_\alpha(p, N-p)$$

$$= 1-\alpha.$$

A loss function justification for (9.1) can be given in frequentist terms analogous to example (2.7). However a somewhat similar approach in Bayesian terms is as follows: In general let $p(\theta) \in \mathbf{P}$ a class of prior densities. Then for each member of $\mathbf{P}$ we obtain a corresponding predictive density for X_{N+1}, namely g and class G assuming it exists. We then calculate

$$K(f, g) = E\left[\log \frac{f(x_{N+1}|\theta)}{g(x_{N+1}|x^{(N)})}\right]$$

where the expectation is over the distribution of X_{N+1} given θ. Next we obtain

$$E[K(f, g)]$$

where now the expectation is over $X_1, \ldots, X_N$ given θ. We then search among the class P and its correspondents G for the prior $p(\theta)$ that minimizes $E[K(f, g)]$. In our particular case if we restrict the class **P** such that the correspondent of each $p(\theta)$ for $\theta = (\mu, \Sigma^{-1})$ is a $g(x_{N+1}|x^{(N)})$ such that g is invariant under translation and nonsingular linear transformations of the sample space, then one can show that the unique solution is

$$p(\mu, \Sigma^{-1}) \propto |\Sigma|^{(p+1)/2}.$$

9.2   Normal linear calibration

Two methods may be available for measuring the same phenomenon. One may be highly accurate but time consuming and expensive while the other may be inexpensive, quicker, and easier to perform but less accurate. An experiment has been run to calibrate the cheaper method in a random way (sometimes called a natural experiment), i.e., samples have been drawn at random from the population yielding N pairs of observations $(y_i, x_i) = d_i$, where d_i stands for the ith data pair and $d^{(N)}$ for the data set. We assume that the calibration is linear and the joint distribution of $(Y, X) = D$ is $N(\mu, \Sigma)$ where

$$\mu = \begin{pmatrix} \mu_1 \\ \mu_2 \end{pmatrix}, \qquad \Sigma = \begin{pmatrix} \sigma_{11} & \sigma_{12} \\ \sigma_{21} & \sigma_{22} \end{pmatrix}$$

noting the simplification in notation from (X_1, X_2) to (X, Y).

In this case Y and X represent the less accurate and the more accurate measurements, respectively. For a future sample value $D_{N+1} = (X_{N+1}, Y_{N+1})$ we observe $Y_{N+1} = z$, the object is to predict the unobserved value, say w, corresponding to X_{N+1}. This is accomplished by calculating the predictive distribution of (X_{N+1}, Y_{N+1}) and then the conditional predictive distribution of X_{N+1} given $Y_{N+1} = z$. From our previous work (9.1) where we assumed a particular prior and $p = 2$ we obtain

$$f(w, z | d^{(N)}) \propto |(N-1)S + \frac{N}{N+1}\begin{pmatrix} w - \bar{x} \\ z - \bar{y} \end{pmatrix}(w - \bar{x}, z - \bar{y})|^{-N/2}.$$

Now clearly both

$$f(z | w, d^{(N)}) \propto f(w, z | d^{(N)})$$

and

$$f(w | z, d^{(N)}) \propto f(w, z | d^{(N)}).$$

For $N\bar{x} = \sum_{i=1}^{N} x_i$, $N\bar{y} = \sum_{i=1}^{N} y_i$ and

$$S = \begin{pmatrix} s_{11} & s_{12} \\ s_{21} & s_{22} \end{pmatrix}$$

where

$$s_{11} = (N-1)^{-1} \sum_{i=1}^{N} (x_i - \bar{x})^2,$$

$$s_{12} = (N-1) \sum_{i=1}^{N} (x_i - \bar{x})(y_i - \bar{y}),$$

$$s_{22} = (N-1) \sum_{i=1}^{N} (y_i - \bar{y})^2,$$

let

$$s_{11 \cdot 2} = s_{11} - \frac{s_{12}^2}{s_{22}}, \qquad \hat{\beta}_{1 \cdot 2} = \frac{s_{12}}{s_{22}}$$

$$s_{22 \cdot 1} = s_{22} - \frac{s_{21}^2}{s_{11}}, \qquad \hat{\beta}_{2 \cdot 1} = \frac{s_{21}}{s_{11}}.$$

Then after some algebra we obtain

$$Y_{N+1}|w \sim S\left[N - 2, \bar{y} + \hat{\beta}_{2 \cdot 1}(w - \bar{x}), (N - 2)s_{22 \cdot 1}\right.$$

$$\left. \times \left[1 + \frac{1}{N} + \frac{(w - \bar{x})^2}{(N - 1)s_{11}}\right]\right].$$

The center of the calibration distribution turns out to be simply the inverse regression of X on Y, since

$$X_{N+1}|z \sim S\left[N - 2, \bar{x} + \hat{\beta}_{1 \cdot 2}(z - \bar{y}), (N - 2)s_{11 \cdot 2}\right.$$

$$\left. \times \left[1 + \frac{1}{N} + \frac{(z - \bar{y})^2}{(N - 1)s_{22}}\right]\right]. \qquad (9.2)$$

There is another kind of calibration experiment that is a designed experiment where the values $x_1, \ldots, x_N$ are selected or controlled and the $y_1, \ldots, y_N$ observed. In a designed experiment there is no sample information on X from the trial itself (e.g., Geisser, 1964), and

$$f(w, z) = f(z|w)f(w)$$

where the density of X_{N+1} must be obtained from information other than the sample. Once ascertained then the conditional predictive or calibrative distribution can be obtained. However if we assume the following:

$$Y|X \sim N\left(\alpha + \beta_{2 \cdot 1}x, \sigma^2\right)$$

$$X_{N+1} \sim N(\eta, \gamma)$$

$$p\left(\alpha, \beta_{2 \cdot 1}, \sigma^2\right) \propto \frac{1}{\sigma^2}$$

for known η and γ then one can obtain excellent numerical approximations to the calibrative distribution of X_{N+1} given z. It is of some interest to note that if $X_i \sim N(\eta, \gamma)$, $i = 1, \ldots, N$ and we assume

$$p(\eta, \gamma) \propto \text{Const.},$$

we obtain the previous calibrative distribution (9.2) (Geisser, 1985a). For discussions of Bayesian calibration, see Dunsmore (1968), Hoadley (1970), and Brown (1982).

9.2.1 Numerical illustration of calibration data

An inexpensive portable analyzer for measuring the number of micrograms of a pollutant per cubic meter of air in samples has become available. The usual accurate method requires bringing the samples back to a laboratory for analysis and is much more costly. Twenty random samples have been collected and are analyzed by both methods. The bivariate data are $(\bar{x}, \bar{y}) = (35.14, 39.55)$ and $(s_{11}, s_{22}, s_{12}) = (1.874, 10.390, 9.650)$ and $(\hat{\beta}_{1 \cdot 2}, \hat{\beta}_{2 \cdot 1}) = (2.99, 1.657)$. The data are plotted in Figure 9.1, along with both regressions Y on X and X on Y and a 0.95 probability prediction interval for X_{N+1} given $Y_{N+1} = z$ over a range of values for z. Thus for a given z, $E(X_{N+1}) = \bar{x} + \beta_{1 \cdot 2}(z - y) = 35.14 + 0.299(z - 39.55)$ is the predicted value of the laboratory analysis given an observed z from the portable analyzer.

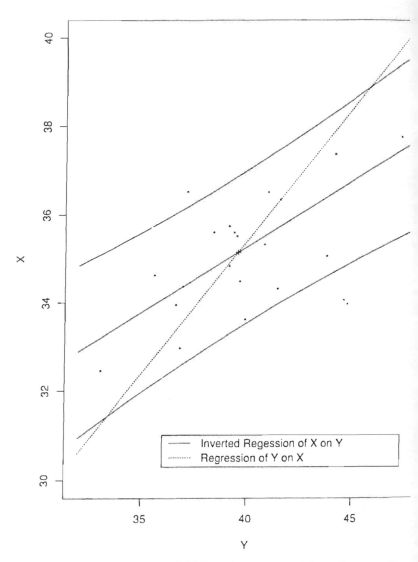

Figure 9.1   *Regressions and 0.95 prediction interval for pollutant calibration*

9.3 Classification

We assume we have a known number, say k, of distinct and labeled populations measured on the same p variables whose distributions are multivariate normal but whose parameters are unknown. One or more unlabeled observations have been or will be generated from these populations. The goal is to classify these observations as to their appropriate populations in an optimal manner. Such procedures are useful in a variety of situations ranging from medical diagnosis to the admission of students to college.

Consider the case where population π_i, $i = 1, \ldots, k$ is assumed to be $N(\mu_i, \Sigma_i)$ and labeled random samples of size N_i (also known as training samples) from $\pi = \pi_i$, $i = 1, \ldots, k$ are available. Now before an unlabeled vector Z is observed we assume the prior probability that it has label π_i is

$$\Pr[\pi = \pi_i] = q_i, \qquad \sum q_i = 1.$$

The object is to determine the posterior probability after observing $Z = z$, and the training sample, namely

$$\Pr[\pi = \pi_i | z, D, \mathbf{q}]$$

where $D = (D_1, \ldots, D_k)$, D_i is the data from π_i, and $\mathbf{q} = (q_1, \ldots, q_k)$ is assumed known for the time being (we shall treat $\mathbf{q}$ unknown subsequently). Now we outline the steps of the calculation (the actual details will be treated in the latter part of the next section). Hence

$$\Pr[\pi = \pi_i | z, D, \mathbf{q}] = q_i f(z | D, \pi_i) \bigg/ \sum_{j=1}^{k} q_j f(z | D, \pi_j)$$

where

$$f(z | D, \pi_i) = \int f(z | \mu_i, \Sigma_i) p(\mu_i, \Sigma_i | D) \, d\mu_i \, d\Sigma_i$$

and $p(\mu_i, \Sigma_i | D)$ is obtained from

$$p(\mathbf{\mu}, \mathbf{\Sigma} | D) \propto p(\mathbf{\mu}, \mathbf{\Sigma}) \prod_{i=1}^{k} f(D_i | \mu_i, \Sigma_i)$$

where $\mathbf{\mu} = (\mu_1, \ldots, \mu_k)$ and $\mathbf{\Sigma} = (\Sigma_1, \ldots, \Sigma_k)$ and $p(\mathbf{\mu}, \mathbf{\Sigma})$ is the prior density of all the parameters. The calculation is made as follows:

$$f(z | \mathbf{\mu}, \mathbf{\Sigma}) = \sum_j q_j f(z | \mu_j, \Sigma_j),$$

$$\Pr[\pi = \pi_i | z, \mathbf{\mu}, \mathbf{\Sigma}, \mathbf{q}] = q_i f(z | \mu_i, \Sigma_i) \Big/ \sum_j q_j f(z | \mu_j, \Sigma_j),$$

$$\int \Pr[\pi = \pi_i | z, \mathbf{\mu}, \mathbf{\Sigma}, \mathbf{q}] f(z | \mathbf{\mu}, \mathbf{\Sigma}) p(\mathbf{\mu}, \mathbf{\Sigma} | D) \, d\mathbf{\mu} \, d\mathbf{\Sigma}$$

$$\propto \Pr[\pi = \pi_i | z, D, \mathbf{q}]. \tag{9.3}$$

For allocation purposes we may choose to assign z to that π_i for which (9.3) is a maximum, if we ignore differential losses in misclassification. We could also divide the observation space of Z into sets of regions $R_1, \ldots, R_k$ where R_i is the set of regions for which $u_i(z) = q_i f(z | D, \pi_i)$ is maximal and use these as allocating regions for future observations. We may also compute "classification errors," based on the predictive distributions, that are in a sense a measure of the discriminatory power of the variables or characteristics. If we let $\Pr\{\pi_j | \pi_i\}$ represent the predictive probability that z has been classified as belonging to π_j when in fact it belongs to π_i, then we obtain

$$\Pr\{\pi_i | \pi_i\} = \int_{R_i} f(z | D, \pi_i) \, dz,$$

$$\Pr\{\pi_j | \pi_i\} = \int_{R_j} f(z | D, \pi_i) \, dz, \qquad i \neq j,$$

$$\Pr\{\pi_i^c | \pi_i\} = 1 - \int_{R_i} f(z | D, \pi_i) \, dz$$

where π_i^c stands for all the populations with the exception of π_i.

Then the predictive probability of a misclassification is

$$\sum_{i=1}^{k} q_i \Pr\{\pi_i^c | \pi_i\} = 1 - \sum_{i=1}^{k} q_i \Pr\{\pi_i | \pi_i\}.$$

Prior to observing Z, the smaller the predictive probability of a misclassification the more confidence we have in the discriminatory variables. However, once Z has been observed and if our interest is only in the particular observed z, the misclassification errors are relatively unimportant, but what is important is the posterior probability that z belongs to π_i. Nevertheless, before any observations are inspected for assignment, the error of classification can be of value in determining whether the addition of new variables or the deletion of old ones is warranted.

If losses of misclassification vary, then this can be incorporated to determine optimal regions that minimize the loss of misclassification. Suppose the cost of misclassifying the label of z as π_j when in fact it is π_i is given as $L(j, i)$ then the expected loss is

$$E(L) = \sum_{i=1}^{k} \sum_{\substack{j=1 \\ j \neq i}}^{k} L(j, i) q_i \Pr(\pi_j | \pi_i).$$

Now

$$\min_{R_1, \ldots, R_k} E(L)$$

requires that z be labeled as π_t if

$$\sum_{\substack{t=1 \\ i \neq t}}^{k} L(t, i) q_i f(z | D, \pi_i) < \sum_{\substack{i=1 \\ i \neq j}}^{k} L(j, i) q_i f(z | D, \pi_i).$$

In many situations the q_is are unknown. First we consider that the sampling situation was such that we have the multinomial density for N_is (where throughout what follows $N_k = N - N_1 - \cdots - N_{k-1}$, and $q_k = 1 - q_1 - \cdots - q_{k-1}$). Thus the likelihood for the observed frequencies in the training sample is

$$L(q_1,\ldots,q_{k-1}) \propto \prod_{j=1}^{k} q_j^{N_j}.$$

If we assume that the prior probability density of the q_is is of the Dirichlet form

$$p(q_1,\ldots,q_{k-1}) \propto \prod_{j=1}^{k} q_j^{\alpha_j},$$

we obtain the posterior density of the q_is,

$$p(q_1,\ldots,q_{k-1}|N_1,\ldots,N_{k-1}) \propto \prod_{j=1}^{k} q_j^{N_j+\alpha_j}.$$

Further

$$p(q_1,\ldots,q_{k-1}|z,N_1,\ldots,N_{k-1}) \propto p(q_1,\ldots,q_{k-1}|N_1,\ldots,N_{k-1})$$
$$\times \sum_{j=1}^{k} q_j f(z|D,\pi_j),$$

from which we obtain the posterior probability no longer conditioned on $\mathbf{q}$,

$$\Pr[\pi = \pi_i|D,z]$$

$$= \int \cdots \int \Pr[\pi = \pi_i|D,z,\mathbf{q})$$

$$\times p(q_1,\ldots,q_{k-1}|z,N_1,\ldots,N_{k-1}) dq_1,\ldots,dq_{k-1}$$

$$= \frac{(N_i+\alpha_i+1)f(z|D,\pi_i)}{\sum_j (N_j+\alpha_j+1)f(z|D,\pi_j)}.$$

In the controlled or designed situation we assume that the N_is were controlled. This is tantamount to assuming that $N_i = 0$ for all

i as regards the posterior distribution of the q_is, resulting in

$$\Pr(\pi = \pi_i | D, z) \propto (\alpha_i + 1) f(z | D, \pi_i).$$

The α_is may be regarded as reflecting previous frequencies or intuitive impressions about the frequencies of the various π_is. If there is neither previous data nor any other kind of prior information, the assumption $\alpha_i = \alpha$ for all i leads to the same result that we would obtain had we assumed that the k populations were all equally likely a priori, i.e., $q_i = 1/k$.

For variations of the case given here and in particular when $\Sigma_i = \Sigma$ for all i see Geisser (1964, 1967, 1982), Geisser and Desu (1968), and Enis and Geisser (1974). These algorithms and others will be given in the next section.

9.3.1 Marginal, joint, and sequential classification

Suppose we wish to classify jointly n independently drawn observations $z_1, \ldots, z_n$, each having prior probability q_i of belonging to π_i. We can then compute the joint predictive density on the hypothesis that $(Z_1 \in \pi_{i_1}, \ldots, Z_n \in \pi_{i_n})$ for $i_j \in (1, \ldots, k)$,

$$f(z_1, \ldots, z_n | D, \pi_{i_1}, \ldots, \pi_{i_n}) = \int p(\mu, \Sigma | D) \prod_{j=1}^{n} f(z_j | \mu_{i_j}, \Sigma_{i_j}) \, d\mu \, d\Sigma.$$

Here we would obtain

$$\Pr\{\pi_{i_1}, \ldots, \pi_{i_n} | D, z_1, \ldots, z_n\} \propto \left(\prod_{j=1}^{n} q_{i_j} \right) f(z_1, \ldots, z_n | D, \pi_{i_1}, \ldots, \pi_{i_n}).$$

It should be noted that because $Z_1, \ldots, Z_n$ are dependent in the predictive distribution the probability of a joint classification will differ from the product of the probabilities of single classification as would be reflected in the case of the previous section. Again, one could now jointly assign $z_1, \ldots, z_n$ by calculating the maximum of the above w.r.t. $\pi_{i_1}, \ldots, \pi_{i_n}$.

Another form of predictive classification would be one wherein allocations need be made as soon as possible, i.e., as soon as Z_i is observed. Hence, if $Z_1, Z_2 \ldots$ are to be observed sequentially, we may wish, when we are ready to observe and classify Z_n, to make our allocation as precise as possible by incorporating the previous observations $z_1, \ldots, z_{n-1}$ into our predictive apparatus. We need now compute the sequential predictive density of Z_n on the hypothesis that it belongs to π_i conditional on the observations D (whose population origins are known), and on the observations $z_1, \ldots, z_{n-1}$ (whose population origins are uncertain). We then obtain the sequential predictive density of Z_n on the hypothesis that it belongs to π_i.

$$f(z_n | D, z_1, \ldots, z_{n-1}, \pi_i) \propto \sum_{i_{n-1}=1}^{k} \cdots \sum_{i_1=1}^{k} q_{i_1}, \ldots, q_{i_{n-1}}$$

$$\times f(z_1, \ldots, z_n | D, \pi_i, \ldots, \pi_{i_{n-1}}, \pi_i),$$

i.e., a mixture of joint predictive densities with Z_n assumed from π_i. Further,

$$\Pr\{\pi = \pi_i | D, z_1, \ldots, z_{n-1}\} \propto q_i f(z_n | D, z_1, \ldots, z_{n-1}, \pi_i).$$

Assignment of z_n can be made to the π_i that maximized the above.

We now apply this to multivariate normal distributions. The usual situation is to assume equal covariance matrices but differing means for the k populations $\pi_1, \ldots, \pi_k$. Hence π_i is represented by an $N(\mu_i, \Sigma)$ distribution with an available training sample $x_{i1}, \ldots, x_{iN_i}$, $i = 1, \ldots, k$. We define

$$\bar{x}_i = N_i^{-1} \sum_{j=1}^{N_i} x_{ij}, \qquad (N_i - 1)S_i = \sum_j (x_{ij} - \bar{x}_i)(x_{ij} - \bar{x}_i)',$$

$$(N-k)S = \sum_i (N_i - 1)S_i, \qquad N = \sum_{i=1}^{k} N_i.$$

Using the prior for $\mu_1, \ldots, \mu_k$ and Σ^{-1},

$$p(\Sigma^{-1}, \mu_1, \ldots, \mu_k) \propto |\Sigma|^{(p+1)/2},$$

we easily obtain, including only relevant constants,

$$
f(z|\bar{x}_i, S, \pi_i) \propto \left(\frac{N_i}{N_i + 1}\right)^{p/2}
$$

$$
\times \left[1 + \frac{N_i(\bar{x}_i - z)'S^{-1}(\bar{x}_i - z)}{(N_i + 1)(N - k)}\right]^{-(N - k + 1)/2},
$$

so that

$$
\Pr[\pi = \pi_i|D, \mathbf{q}] \propto q_i f(z|\bar{x}_i, S, \pi_i). \tag{9.4}
$$

When this is used for each new z to be classified, we refer to this as marginal classification.

We now assume that we need jointly allocate n new vector observations $Z_1, \ldots, Z_n$. Letting $\mathbf{Z}_i$ represent the matrix of n_i observations assumed from π_i, with $n = \sum_{i=1}^{k} n_i$, we obtain

$$
f(\mathbf{z}_1, \ldots, \mathbf{z}_k|D, \pi_1, \ldots, \pi_k)
$$

$$
\propto \left(\prod_{i=1}^{k} \frac{N_i}{N_i + n_i}\right)^{p/2}
$$

$$
\times \left|(N - k)S + \prod_{i=1}^{k}(\mathbf{z}_i - \bar{x}_i e_i')\Omega_i(\mathbf{z}_i - \bar{x}_i e_i')'\right|^{-(N + n - k)/2}
$$

where $\Omega_i = I + N_i^{-1} e_i e_i'$ and $e_i' = (1, \ldots, 1)$ is of dimension n_i. Hence for the joint probability that $\mathbf{Z}_i \in \pi_i$ we obtain, changing the notation once more for convenience

$$
\Pr[\pi_1, \ldots, \pi_k|D, \mathbf{z}_1, \ldots, \mathbf{z}_k, \mathbf{q}]
$$

$$
\propto \left(\prod_{i=1}^{k} q_i^{n_i}\right) f(\mathbf{z}_1, \ldots, \mathbf{z}_k|D, \pi_1, \ldots, \pi_k).
$$

The observations may, in many instances, be sequentially obtained and for compelling reasons allocations (diagnoses) need to be made immediately.

Let $z^{(n-1)} = (z_1, \ldots, z_{n-1})$ and Σ' stand for the sum over all assignments of $z_1, \ldots, z_{n-1}$ to $\mathbf{z}_1, \ldots, \mathbf{z}_k$ with z_n always assigned to z_i and then summed over all partitions of n such that $\sum_{j-1}^{k} n_j = n$, $n_j \geq 0$, $j \neq i$ and $n_i \geq 1$. Then the probability that $Z_n \in \pi_i$ is

$$\Pr\left[\pi = \pi_i | D, z^{(n-1)}, \mathbf{q}\right]$$

$$\alpha \Sigma' \left(\prod_{j=1}^{k} q_j^{n_j}\right) \left(\prod_{j-1}^{k} \frac{N_j}{N_j + n_j}\right)^{p/2}$$

$$\times \left| (N-k)S + \sum_{i-1}^{k} (\mathbf{z}_i - \bar{x}_i e_i') \Omega_i (\mathbf{z}_i - \bar{x}_i e_i')' \right|^{-(N+n-k)/2}$$

for $n = 2, 3 \ldots$ and $\Omega_i = I + e_i e_i'$.

A second case that is also easily managed is the unequal covariance matrix situation. Here π_i is represented by an $N(\mu_i, \Sigma_i)$ distribution $i = 1, \ldots, k$.

Using the same training sample notation as previously and the usual prior

$$p\left(\mu_1, \ldots, \mu_k, \Sigma_1^{-1}, \ldots, \Sigma_k^{-1}\right) \alpha \prod_{i=1}^{k} |\Sigma_i|^{(p+1)/2}$$

we obtain

$$f(z | \bar{x}_i, S_i \pi_i)$$

$$\alpha \left(\frac{N_i}{N_i + 1}\right)^{p/2} \frac{\Gamma(N_i/2)[1 + \{[N_i(\bar{x}_i - z)'S_i^{-1}(\bar{x}_i - z)]/(N_i^2 - 1)\}]^{-N_i/2}}{\Gamma[(N_i - p)/2]|(N_i - 1)S_i|^{1/2}}$$

$$(9.5)$$

as the predictive density of the observation to be allocated. Hence the posterioi probability of z belonging to π_i is

$$\Pr\left[\pi = \pi_i | D, \mathbf{q}\right] \alpha q_i f(z | \bar{x}_i, S_i, \pi_i).$$

When all future observations are classified using the above we term this as marginal classification.

For the joint classification of $Z_1, \ldots, Z_n$ we obtain for the joint probability

$$\Pr[\pi_1, \ldots, \pi_k | D, \mathbf{q}] \propto \prod_{i=1}^{k} q_i^{n_i} d(\mathbf{Z}_i | \bar{x}_i e'_i, \Omega_i, S_i, N_i - 1, n_i, p)$$

where

$$d(Y | \Lambda, \Omega, A, M, m, p)$$
$$= \frac{(2\pi)^{-pm/2} K(p, M) |MA|^{M/2} |\Omega|^{p/2}}{K(p, M+m) |MA + (Y - \Lambda) \Omega (Y - \Lambda)'|^{(M+m)/2}} \qquad (9.6)$$

for $M \geq p$, $m \geq 1$, A is $p \times p$ and positive definite, Ω is $m \times m$ and positive definite, Y and Λ are $p \times m$, in addition (9.6) is defined as 1 for $m = 0$, and

$$K^{-1}(p, \nu) = 2^{p\nu/2} \pi^{[p(p-1)]/4} \prod_{j=1}^{p} \Gamma\left(\frac{\nu + 1 - j}{2}\right).$$

For sequential allocation we obtain for $n = 2, 3 \ldots$ for the probability that $z_n \in \pi_i$,

$$\Pr\{\pi = \pi_i | D, z^{(n-1)}, \mathbf{q}\} \propto \sum' \prod_{j=1}^{k} q_j^{n_j} d(\mathbf{z}_j | \bar{x}_j e'_j, \Omega_j, S_j, N_j - 1, n_j, p).$$

where Σ' stands for the sum over all partitions of $n = \sum_{j=1}^{k} n_j$ of assignments of $z_1, \ldots, z_{n-1}$ to $\mathbf{z}_1, \ldots, \mathbf{z}_k$ where $\mathbf{z}_j$ contains $n_j \geq 0$ of the $z_1, \ldots, z_{n-1}$ but z_n is always assigned to $\mathbf{z}_i$ and $n_i \geq 1$. For other cases see Geisser (1966).

Although the three methods (marginal, joint, and sequential) will generally not appreciably differ in the classification of the observations, one can concoct examples where all three methods will yield different answers for two future observations (Geisser, 1966).

When classification between $k = 2$ populations with equal covariance matrices is at issue and $N_1 = N_2 = N$, $q_1 = q_2$, there is a simplification. This results from the fact that

$$\Pr\left[\pi = \pi_1 | D, \tfrac{1}{2}\right] > \Pr\left[\pi = \pi_2 | D, \tfrac{1}{2}\right]$$

if and only if

$$V = \left[z - \tfrac{1}{2}(\bar{x}_1 + \bar{x}_2)\right]' S^{-1}(\bar{x}_1 - \bar{x}_2) > 0.$$

V is termed the linear discriminant. (Actually, V is often used in frequentist contexts even when $N_1 \neq N_2$.) To ascertain the predictive error of classification we can calculate

$$\Pr\left[V < 0 | z \in \pi_1\right] = E\left\{\Pr\left[V < 0 | z \in \pi_1, \mu_1, \mu_2, \Sigma^{-1}\right]\right\}$$

where the expectation is over the joint posterior distribution of μ_1, μ_2, and Σ^{-1} namely

$$p\left(\mu_1, \mu_2, \Sigma^{-1} | D\right)$$

$$\alpha \; |\Sigma^{-1}|^{(\nu-p+1)/2} e^{-(1/2)\text{tr} \, \Sigma^{-1}\{\nu S + [N_1 N_2 /(N_1 + N_2)](\bar{x}_1 - \bar{x}_2)(\bar{x}_1 - \bar{x}_2)'\}}$$

where $\nu = N_1 + N_2 - 2$. Since V conditional on μ_1, μ_2, and Σ^{-1} is normally distributed, the integration over μ_1, μ_2, and Σ^{-1} yields

$$\Pr\left[V < 0 | z \in \pi_1\right] = \Pr\left\{t_{\nu+1-p} < -\frac{1}{2}\left[\frac{QN_1(\nu+1-p)}{\nu(N_1+1)}\right]^{1/2}\right\}$$

where

$$Q = (\bar{x}_1 - \bar{x}_2)' S^{-1}(\bar{x}_1 - \bar{x}_2)$$

and $t_{\nu+1-p}$ is the Student's t with $\nu + 1 - p$ degrees of freedom. For the other error we calculate

$$\Pr\left[V > 0 | z \in \pi_2\right] = \Pr\left\{t_{\nu+1-p} > \frac{1}{2}\left[\frac{QN_2(\nu+1-p)}{\nu(N_2+1)}\right]^{1/2}\right\}.$$

Setting $N_1 = N_2 = N$ will result precisely in the same classification as the posterior probability ratio. Further

$$\Pr[V > 0 | z \in \pi_2] = \Pr[V < 0 | z \in \pi_1] = P,$$

so that the predictive probability of correct classification using the linear discriminant is just $1 - P$. The interpretation is that if this known discriminant is used on all future observations and $q_1 = q_2$ then $1 - P$ will be the fraction of them correctly classified.

When $N_1 \neq N_2$, classification from the posterior probabilities and the linear discriminant is no longer the same although the difference may be negligible in many cases.

9.3.2 Numerical illustration of normal classification

Andrews and Hertzberg (1985) presented 18 skull measurements on 3 species of male and female kangaroos. We have chosen a set of 24 female and 25 male M. *giganteus* and 2 measurements, basilar length and zygomatic width. Ten males and 10 females were randomly chosen on these two variables as training samples. These samples are presented in Table 9.1. The problem is to classify future M. *giganteus* skulls as to sex.

The training sample means and covariance matrices, in an obvious notation, are $\bar{x}'_M = (1558.8, 885.9)$, $\bar{x}'_F = (1413.2, 833.2)$.

$$S_M = \begin{bmatrix} 33353.29 & 13394.64 \\ & 5733.43 \end{bmatrix}$$

$$S_F = \begin{bmatrix} 18344.84 & 8171.51 \\ & 3799.73 \end{bmatrix}$$

and the pooled covariance matrix

$$S = \begin{bmatrix} 25849.07 & 10783.08 \\ & 4766.58 \end{bmatrix}.$$

Assuming $q_M = q_F = 1/2$, $N_M = N_F = 10$, $k = 2$, homogeneous covariance matrices, and using the marginal method for classification

$$\Pr[\pi = \pi_M | D] > \Pr[\pi = \pi_F | D]$$

if

$$V = \left[z - \tfrac{1}{2}(\bar{x}_M + \bar{x}_F) \right]' S^{-1} (\bar{x}_M - \bar{x}_F) > 0.$$

Since

$$Q = (\bar{x}_M - \bar{x}_F)' S^{-1} (\bar{x}_M - \bar{x}_F) = 1.061,$$

the probability of misclassification for males is

$$\Pr\left[t_{17} \leq -\frac{1}{2} \left(\frac{1.061 \times 10 \times 17}{18 \times 11} \right)^{1/2} \right] = 0.32.$$

The same probability is attained for females so that the overall probability of correct classification is 0.68.

If on the other hand we assume that the covariance matrices are not homogenous, then using (9.5)

$$\Pr(\pi = \pi_M | D) > \Pr(\pi = \pi_F | D)$$

if

$$\left[\frac{1 + (10/99)(\bar{x}_F - z)' S_F^{-1}(\bar{x}_F - z)}{1 + (10/90)(\bar{x}_M - z)' S_M^{-1}(\bar{x}_M - z)} \right]^5 \frac{|S_F|^{1/2}}{|S_M|^{1/2}} > 1.$$

Table 9.1　*Male and Female Training Samples on Basilar Length X_1 and Zygomatic Width X_2 and Classification into Male or Female via E = Pooled Covariance Matrices or U = Separate Covariance Matrices*

Males				Females			
X_1	X_2	U	E	X_1	X_2	U	E
1439	824	M	M	1464	848	F	F
1413	823	F	F	1262	760	F	F
1490	897	F	F	1112	702	F	F
1612	921	F	M	1414	853	F	F
1388	805	F	F	1427	823	F	M
1840	984	M	M	1423	839	F	F
1294	780	F	F	1462	873	F	F
1740	977	M	M	1440	832	F	F
1768	968	M	M	1570	894	M	M
1604	880	M	M	1558	908	F	F

Table 9.2 *Male and Female Future Samples and Their Classification by the Two Methods*

Males				Females			
X_1	X_2	U	E	X_1	X_2	U	E
1312	782	F	F	1373	828	F	F
1378	778	M	M	1415	859	F	F
1315	801	F	F	1374	833	F	F
1090	673	F	F	1382	803	F	F
1377	812	F	F	1575	903	F	M
1296	759	F	F	1559	920	F	F
1470	856	F	F	1546	914	F	F
1575	905	F	M	1512	885	F	F
1717	960	M	M	1400	878	M	F
1587	910	F	M	1491	875	F	F
1630	902	M	M	1530	910	F	F
1552	852	M	M	1607	911	M	M
1595	904	M	M	1589	911	F	M
1846	1013	M	M	1548	907	F	F
1702	947	M	M				

This implies that z will be classified as female if it lies within an ellipse (quadratic) and male otherwise.

The classification probabilities can be estimated fairly accurately by generating 2500 samples from the appropriate bivariate student distributions. This yielded a misclassification rate for males as 0.38 and for females 0.19 or an overall probability of correct classification of 0.72.

Finally the additional 15 male and 14 female kangaroo skull values were classified using both approaches in Table 9.2. The

Table 9.3 *Fraction of Kangaroos Correctly Classified for the Two Methods*

	Method	
	E	U
Males		
Training sample	6/10	5/10
Future sample	9/15	7/15
Females		
Training sample	8/10	9/10
Future sample	11/14	12/14

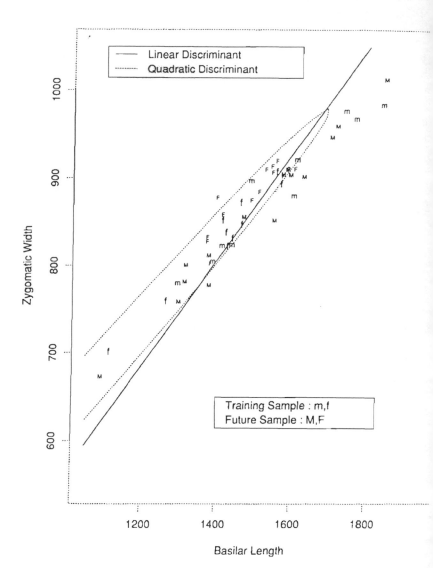

Figure 9.2   *Linear and quadratic discriminants based on training samples*
(*m, f*) *for* M. giganteus *kangaroos.*

original training samples were also classified by these methods in Table 9.1. A summary of the results is presented in Table 9.3. A graph of the two variables for the 29 kangaroos is given in Figure 9.2, which includes the two methods of discriminating based on the training samples. We note that the relative frequency of correctly classified kangaroos that are not in the training sample is 0.69 and 0.66 for the equal and unequal covariance cases, respectively. This is rather close to probabilities calculated for them, namely 0.68 and 0.72.

9.4 The general linear model

9.4.1 *Joint prediction of* $X_{(M)}$

Consider the situation where joint prediction of M p-dimensional vectors is at issue $X_{(M)} = (X_{N+1}, \ldots, X_{N+M})$ where X_{N+i} $i = 1, \ldots, M$ is a p-dimensional $N(\mu, \Sigma)$ variate. Here we can show, in a fashion similar to our previous work, using the matrix identity

$$n(A - C)(A - C)' + m(B - C)(B - C)'$$
$$= (n + m)(D - C)(D - C)' + \frac{nm}{n + m}(A - B)(A - B)'$$

where A, B, and C are $p \times q$ matrices, n and m scalars such that $n + m \neq 0$, and $D = (n + m)^{-1}(nA + mB)$, that

$$f(x_{(M)}|\bar{x}, S)$$
$$= \frac{(2\pi)^{-pM/2} K(p, N - 1)|(N - 1)S|^{(N-1)/2}|\Omega|^{p/2}}{K(p, N - 1 + M)|(N - 1)S + (x_{(M)} - \bar{x}e')\Omega(x_{(M)} - \bar{x}e')'|^{(N+M-1)/2}}$$

(9.7)

where $\Omega = I + N^{-1}ee'$ and $e' = (1, \ldots, 1)$ is of dimension M.

More generally suppose the linear model (changing notation to conform with the way the multivariate linear model is usually presented)

$$Y_j = \beta X_j + e_j$$

obtains, where X_j is a $q \times 1$ vector of known covariates, β is a $p \times q$ regression matrix, and e_j is a p-dimensional $N(0, \Sigma)$ variates where $j = 1, \ldots, N$. Further let $N \geq p + q$ and

$$X = (X_1, \ldots, X_N)$$

be of rank q, $Y = (Y_1, \ldots, Y_N)$ and $e = (e_1, \ldots, e_N)$. Then in matrix notation

$$Y = \beta X + e.$$

(The reader should be aware that the notation here does not directly conform to the one-dimensional case for univariate regression of Chapter 3, Section 9 unless one here sets $p = 1$ and $q = p$ and transposes Y, X, and β.)

The joint density is then written as

$$f(y|X, \beta, \Sigma) = \frac{|\Sigma|^{-N/2}}{(2\pi)^{pN/2}} e^{-(1/2)\mathrm{tr}\, \Sigma^{-1}(y - \beta X)(y - \beta X)'}.$$

Using

$$p(\beta, \Sigma^{-1}) \propto |\Sigma|^{(p+1)/2}$$

and the algebraic identity

$$(y - \beta X)(y - \beta X)' = (y - \hat{\beta} X)(y - \hat{\beta} X)'$$
$$+ (\beta - \hat{\beta}) XX'(\beta - \hat{\beta})'$$

where $\hat{\beta} = yX'(XX')^{-1}$ then

$$p(\beta, \Sigma^{-1}|y, X) \propto |\Sigma^{-1}|^{(N-p-1)/2} e^{-(1/2)\mathrm{tr}\, \Sigma^{-1}[A + (\beta - \hat{\beta})XX'(\beta - \hat{\beta})']}$$

where

$$A = (y - \hat{\beta} X)(y - \hat{\beta} X)'.$$

Now suppose that $Z = (Z_1, \ldots, Z_M)$ is a set of future observations such that

$$Z = \beta W + e,$$

where $W = (W_1, \ldots, W_M)$ is $q \times M$ and $e = (e_1, \ldots, e_M)$ and $e_1, \ldots, e_M$ are independently distributed as before. Then the predictive density of the set Z is

$$f(z|y) = \int f(z|W, \beta, \Sigma^{-1}) p(\Sigma^{-1}, \beta|y, X) \, d\beta \, d\Sigma^{-1}$$

$$\alpha \int |\Sigma^{-1}|^{(M+N-p-1)/2} e^{-(1/2)\operatorname{tr} \Sigma^{-1}(v - \beta U)(v - \beta U)'} d\beta \, d\Sigma^{-1}$$

where

$$v = (y, z), \qquad U = (X, W).$$

Letting $\hat{\beta}_0 = vU'(UU')^{-1}$ it is easy to see that

$$f(z|y) \alpha |(v - \hat{\beta}_0 U)(v - \hat{\beta}_0 U)'|^{-(N+M-q)/2}.$$

Next the algebraic identity

$$\left(V - \hat{\beta}_0 U\right)\left(V - \hat{\beta}_0 U\right)' = A + (Z - \hat{\beta} W)$$
$$\times \left[I - W'(UU')^{-1}W\right](Z - \hat{\beta} W)'$$

can be established. Hence, recovering the constants

$$f(z|y) = (2\pi)^{-pM/2} K(p, N-q)|A|^{(N-q)/2}|I - W'(UU')^{-1}W|^{p/2}$$
$$\div \{ K(p, N+M-q)|A + (z - \beta W)[I - W'(UU')^{-1}W]$$
$$\times (z - \beta W)'|^{(N+M-q)/2} \}.$$

Further setting

$$0 < \frac{|A|}{|A + (Z - \hat{\beta} W)[I - W'(UU')^{-1}W](Z - \hat{\beta} W)'|} = U_{p, M, N-q},$$

which is bounded above by 1, and computing its moments from

the predictive density we find that $U_{p,M,N-q}$ has the same moments as a product of p independent beta variates say $X_1, \ldots, X_p$ where X_j has density

$$f\left(x \mid \frac{N-q+1-j}{2}, \frac{M}{2}\right) \propto x^{(N-q-1-j)/2}(1-x)^{(M/2)-1} \qquad 0 < x < 1$$

$j = 1, \ldots, p$. Hence $U_{p,M,N-q}$ is distributed as that product of beta variates. In particular, if we are predicting a single future observation, i.e., $M = 1$, then a p-dimensional predictive ellipsoid can be obtained from the above since

$$\frac{N-q-p+1}{p}\left[1 - W_1'(UU')^{-1}W_1\right]\left(Z_1 - \hat{\beta}W_1\right)' A^{-1}\left(Z_1 - \hat{\beta}W_1\right)$$

$$\sim F(p, N-q-p+1)$$

so that

$$\Pr\left[\left(Z_1 - \hat{\beta}W_1\right)' A^{-1}\left(Z_1 - \hat{\beta}W_1\right) \le \frac{p(N-q-p+1)^{-1}}{\left[1 - W_1'(UU')^{-1}W_1\right]}F_\alpha\right]$$

$$= 1 - \alpha = \Pr[F \le F_\alpha]$$

(Geisser, 1965).

9.5   The growth curve model

In this model

$$\beta = H\tau$$

where H is a $p \times m$ known matrix of rank $m < p$ and τ is an unknown $m \times q$ matrix. We use

$$p(\tau, \Sigma^{-1}) \propto |\Sigma|^{(p+1)/2}$$

for the prior density. Hence the posterior density of τ and Σ^{-1} is

$$p(\tau, \Sigma^{-1}|y) \propto |\Sigma^{-1}|^{(N-p-1)/2} e^{-(1/2)\mathrm{tr}\, \Sigma^{-1}(y-H\tau X)(y-H\tau X)'}.$$

Now for the set of M future vectors $(Z_1, \ldots, Z_M) = Z$ such that

$$E(Z) = H\tau F$$

where F is a known $q \times M$ matrix, and

$$f(z|\tau, \Sigma^{-1}) \propto |\Sigma^{-1}|^{(M-p-1)/2} e^{-(1/2)\mathrm{tr}\, \Sigma^{-1}(z-H\tau F)(z-H\tau F)'}$$

then

$$f(z, \tau, \Sigma^{-1}|y)$$
$$\propto |\Sigma^{-1}|^{(N+M-p-1)/2} e^{-(1/2)\mathrm{tr}\, \Sigma^{-1}[(y,z)-H\tau(X,F)][(y,z)-H\tau(X,F)]'}.$$

After integrating out Σ^{-1} and τ, we shall need the following easily proven identity:

If $D_{p \times q}$ and $E_{p \times (p-q)}$ are of ranks q and $p - q$, respectively, such that $E'D = 0$, a null matrix and $S_{p \times p} = S'$ is positive definite, then

$$S^{-1} - S^{-1}D(D'S^{-1}D)^{-1}D'S^{-1} = E(E'SE)^{-1}E'.$$

This identity and further matrix manipulation will result in

$$f(z|y) \propto |G|^{m/2}|U'[yy' + (z - H\hat{\tau}F)(z - H\hat{\tau}F)']U|^{(N+M-m)/2}$$
$$\times |I + G(z - H\hat{\tau}F)'A^{-1}H(H'A^{-1}H)^{-1}$$
$$\times H'A^{-1}(z - H\hat{\tau}F)|^{-(N+M-q)/2} \tag{9.8}$$

where

$$G = \left[I - F'(CC')^{-1}F\right]^{-1}$$
$$+ \left[z - yX'(XX')^{-1}F\right]U(U'AU)^{-1}U'\left[z - yX'(XX')^{-1}F\right]',$$

U is any $p \times (p - m)$ matrix of rank $p - m$ such that $U'H = 0$, $C = (X, F)$,

$$\hat{\tau} = (H'A^{-1}H)^{-1} H'A^{-1}yX'(XX')^{-1}$$

and

$$A = y\left[I - X'(XX')^{-1}X\right]y',$$

(Geisser, 1970). While it is not difficult to obtain

$$E(Z) = H\hat{\tau}F$$

$$\mathrm{Cov}(Z) = (N - p - 1)^{-1}(y - H\hat{\tau}X)(y - H\hat{\tau}X)' \otimes I_M$$
$$+ (N - q - m - 1)H(H'A^{-1}H)^{-1}$$
$$\times H' \otimes \left[(N - p - 1)^{-1}(\mathrm{tr}\, BX'X)I_M + F'BF\right]$$

where

$$B = (XX')^{-1} + T'(U'AU)^{-1}T$$
$$T = U'yH'(HH')^{-1},$$

and the direct product of the $p \times p$ and $M \times M$ matrices C and D, respectively, is

$$C \otimes D = \{c_{ij}D\} = \begin{bmatrix} c_{11}D & \cdots & c_{1p}D \\ \vdots & & \\ c_{p1}D & \cdots & c_{pp}D \end{bmatrix},$$

the distribution of Z is not in a form that is readily accessible to the computation of a predictive region on Z. Approximating it by a multivariate normal random matrix with mean and covariance as given above may be adequate for large samples. For Z partitioned into two sets, say $Z = (Z_{(1)}, Z_{(2)})$ one observed and one to be predicted, one then would calculate from (9.5) $f(z_{(2)}|z_{(1)}, y)$ (see Lee and Geisser, 1972, 1975). This would be useful say in the case of single vector where some of the components had also been observed and prediction was required for the unobserved components. For a review of Bayesian growth curve prediction see Geisser (1980a), and for applications see Lee and Geisser (1975).

9.6 Discordancy testing

In Section 9.1 we showed that

$$(X_{N+1} - \bar{x})' S^{-1}(X_{N+1} - \bar{x}) \sim \frac{p(N^2 - 1)}{N(N-p)} F(p, N-p).$$

Hence let

$$(X_i - \bar{x}_{(i)})' S_{(i)}^{-1}(X_i - \bar{x}_{(i)}) = Q_i$$

be a simple diagnostic for discordancy where $\bar{x}_{(i)}$ and $S_{(i)}$ are the sample mean and covariance matrix based on all but x_i. Then a CPD test for x_C as the most discordant observation can be based on

$$\Pr[Q_C > q_C | Q_C > q_{C-1}] = \frac{1 - F(q_C | p, N - 1 - p)}{1 - F(q_{C-1} | p, N - 1 - p)}$$

where q_C and q_{C-1} are the largest and second largest observed values of the diagnostic Q_i associated with x_C and x_{C-1}.

For the general linear model we could let

$$\frac{|A_{(i)}|}{|A_{(i)} + (Y_i - \hat{\beta} X_i)[I - X_i'(XX')^{-1} X_i](Y_i - \hat{\beta} X_i)'|} = U_i$$

which has a $U_{p,1,N-1-q}$ distribution. Hence this results in

$$\Pr[U_C < u_C | U_C < u_{C-1}] = \frac{G(u_C)}{G(u_{C-1})}$$

where G is the distribution function of $U_{p,1,N-1-q}$, because here the smaller U_i, the more discordant y_i.

9.6.1   *Numerical illustration using the kangaroo data*

Among all 25 male kangaroos we calculate $Q_i(x_1, x_2)$ and we obtain

$$Q_C(1490, 897) = 9.56$$
$$Q_{C-1}(1090, 673) = 7.72.$$

Hence using an F variate with 2 and 22 degrees of freedom,

$$\text{Prob}(Q_C > 9.56 | Q_{C-1} > 7.72) = 0.36.$$

Among all 24 female kangaroos

$$Q_C(1112, 702) = 17.72$$
$$Q_{C-1}(1400, 878) = 12.40.$$

Here using an F variate with 2 and 21 degrees of freedom,

$$\Pr(Q_C > 17.72 | Q_{C-1} > 12.40) = 0.11.$$

It would appear that there is little or no cause for concern about discordancies in these data sets.

References

Andrews, D. F., and Hertzberg, A. M. (1985). *Data: A Collection of Problems from Many Fields for the Student and Research Worker.* Berlin: Springer-Verlag.

Box, G. E. P., and Tiao, G. C. (1973). *Bayesian Inference in Statistical Analysis.* Reading, MA: Addison-Wesley.

Brown, P. J. (1982). Multivariate calibration. *Journal of the Royal Statistical Society B* **44**, 287–321.

Dunsmore, I. R. (1968). A Bayesian approach to calibration. *Journal of the Royal Statistical Society B* **30**, 396–405.

Enis, P., and Geisser, S. (1974). Optimal predictive linear discrimination. *Annals of Statistics* **2**(2), 403–410.

Geisser, S. (1964). Posterior odds for multivariate normal classification. *Journal of the Royal Statistical Society B* **1**, 69–76.

Geisser, S. (1965). Bayesian estimation in multivariate analysis. *Annals of Mathematical Statistics* **56**, 150–159.

Geisser, S. (1966). Predictive discrimination. In *Multivariate Analysis,* P. Krishnaiah (ed.). Academic Press, New York, 149–163.

Geisser, S. (1967). Estimation associated with linear discriminants. *Annals of Mathematical Statistics* **38**, 807–817.

Geisser, S. (1970). Bayesian analysis of growth curves. *Sankhya A* **32**, 53–64.

Geisser, S. (1980a). Growth curve analysis. In *Handbook of Statistics,* Vol. I, P. R. Krishnaiah (ed.). North-Holland, 89–115.

Geisser, S. (1985a). On the predicting of observables: A selective update. In *Bayesian Statistics 2,* J. M. Bernardo et al., (eds.). Valencia, Spain: University Press, 203–230.

Geisser, S., and Cornfield, J. (1963). Posterior distributions for multivariate normal parameters. *Journal of the Royal Statistical Society B* **25**, 368–376.

Geisser, S., and Desu, M. M. (1968). Predictive zero-mean uniform discrimination. *Biometrika* **55**, 519–524.

Hoadley, A. B. (1970). A Bayesian look at inverse linear regression. *Journal of the American Statistical Association* η65, 356–369.

Lee, J. C., and Geisser, S. (1972). Growth curve prediction. *Sankhya A* **34**, 393–412.

Lee, J. C., and Geisser, S. (1975). Applications of growth curve prediction. *Sankhya A* **37**, 239–256.

CHAPTER 10

Interim analysis and sampling curtailment

Many experiments consist of a series of independent observations on subjects or units that, for one reason or another, enter the experiment for treatment sporadically. Often a fixed sample size is agreed on as the minimum necessary to make an inference or decision concerning the treatment. In a frequentist context, the agreed-on sample size will generally depend on the power and size of the test of the hypothesis. This results in the application of a test at a certain α level. In the Bayesian context, although a predetermined sample size is not a determining constituent for computing the posterior odds of one hypothesis versus another, planning for the costs and the administration of an experiment may lead to a determination of sample size prior to embarking on a trial. In any event where experimental procedures are costly, it is of great interest for the investigator to know whether to continue testing a new treatment, drug, or therapy after partially completing an experiment.

10.1 Bayesian background

In many instances, we might assume that a laboratory or a regulatory agency requires that for a new therapy a minimum number of observations, S, must be in hand before an evaluation of the effectiveness of the agent is to be made. Ostensibly if interim analyses of the agent are to be conducted at various specified numbers of observations there is a complete decision theoretic approach that can be applied. This would require the following ingredients:

1. There are two possible decisions to be made at any interim point, namely $\bar{d}_1$, the agent is insufficiently promising to continue the trial, and c, the agent is sufficiently promising to continue sampling. Once the minimum sample size S has been achieved then three decisions are available, the former $\bar{d}_1$ and c plus d_1; the new agent is effective.
2. Losses are specified for correct and incorrect decisions.
3. Costs in time and/or money are rigorously ascertainable.
4. Appropriate prior distributions for the unknown parameters can be assumed or elicited.

Assuming further a known horizon at or beyond S, for the sampling, the problem can be solved theoretically at least, by backward induction (Bellman, 1957). Numerical solutions for similar problems have been given (e.g., Lindley and Barnett, 1965).

Aside from the possible complexity of an exact solution, experimenters would rather not specify the exact interim points, nor are costs, losses, and prior distributions readily available. It is also not clear who should specify the losses, the regulatory agency, the experimenters, or the patients being used. Hence it is not surprising that decision theoretic Bayesian sequential procedures have rarely if ever been implemented.

What we shall advocate here is an informal and certainly more realistic approach to Bayesian interim analyses (Geisser, 1992b). First an interim analysis can be made at any time say at $N \leq S$ observations for testing the efficacy of the agent. For $N < S$ we will only conclude whether the agent is sufficiently promising to continue or abandon the trial. Once S has been equalled or exceeded we can decide that the agent is at least as effective, or not as effective as some standard or still continue sampling.

We shall assume that a prior density can be elicited or some noninformative one can be agreed on. Further we require that at S observations, the probability that a quantity reflecting the effectiveness of the agent is in some set, exceeds some prescribed value that has taken into account losses for correct and incorrect decisions.

10.2 The method

We set the problem to be that a parameter or set of parameters $\theta \in \Theta$ is indicative of the effectiveness of the treatment or agent if $\theta \in \Theta_1$ a subset of Θ. Therefore if the posterior probability is such that

$$\Pr\left[\theta \in \Theta_1 | x^{(S)}\right] \geq p, \tag{10.1}$$

we decide d_1, i.e., the treatment is effective. Otherwise we decide $\bar{d}_1$, that it is ineffective or not sufficiently effective or even withhold the decision.

Now suppose we stop at some unpremeditated sample size N for an interim analysis. At this point we shall be interested in whether to continue or terminate the trial. We choose an M such that $N + M \geq S$ and determine what we would expect to happen if we continued the trial. For $N + M \geq S$ denote, as previously,

$$X^{(N+M)} = \left(X^{(N)}, X_{(M)} \right)$$

$$X^{(N)} = (X_1, \ldots, X_N)$$

$$X_{(M)} = (X_{N+1}, \ldots, X_{N+M}).$$

In light of (10.1), the decision to be made is that $\theta \in \Theta_1$ will depend on $T(X^{(N+M)})$. The latter could, of course, be $X^{(N+M)}$ itself. Assume $T(X^{(N+M)}) \in R$. If $T(X^{(N+M)}) \in R_1, R_1 \cup \bar{R}_1 = R$ [where R_1 is determined to be that region for T such that (10.1) holds] we decide d_1. Otherwise we decide $\bar{d}_1$ or perhaps withhold a decision. Of course $\bar{R}_1$ can be split into $\bar{R}_{11}$ and $\bar{R}_{12}$ to reflect that the treatment is ineffective or still in doubt with further sampling a possibility. We assume that $X^{(N)} = x^{(N)}$ has been observed and that we can calculate the predictive distribution function

$$F_{X_{(M)}}\left(x_{(M)} | x^{(N)} \right).$$

It is clear conceptually that we can now calculate the conditional predictive distribution of T, $F_T(t | x^{(N)})$ and subsequently require

that

$$P = \Pr\left[T \in R_1 | x^{(N)} \right] = \Pr\left[\Pr\left(\theta \in \Theta_1 | x^{(N)}, X_{(M)} \right) \geq p \right],$$

where $X_{(M)}$ is random. On the one hand, if P is small, a continuation of the experiment is highly unlikely to lead to d_1 and we may very well decide to abandon the experiment. On the other hand, if P is sufficiently large we are encouraged to continue the experiment. In other instances a prescribed sample size S may yield an equivocal inference, e.g., there appears to be some tendency for the new therapy to yield a better result than the standard but the posterior odds have not reached a required value. In such a case one might want to determine the probability of at least attaining the required posterior odds for each of various additional sample sizes.

10.3 Predictive applications to parametric testing

Here we give a series of examples applying the previous ideas to the testing of parameters that would reflect a normative evaluation in regard to meaningful functions of future observations such as the average or a proportion of a very large number of them.

Example 10.1. Suppose we have a series of binary variates $X_1, X_2, \ldots$ that are independent conditional on $\theta = \Pr(X_i = 1) = 1 - \Pr(X_i = 0)$ where $X_i = 1$ is a successful outcome of therapy and $X_i = 0$ is a failure.

Suppose a fixed sample size experiment of size $N + M$ is used to test the hypothesis $\theta > a \geq 0$. Let us assume the following criterion; if $Y = \sum_1^{N+M} X_i \geq A$ the agent is to be declared effective. Suppose the experiment was performed sequentially and we had N observations already in hand and we wanted to decide whether it was worthwhile going on until the end of the trial noting that t out of N were successes. We could compute, for the given N and t, the predictive probability of S successes in $N + M$ trials or equivalently R successes the next M trials

$$P = \Pr[Y \geq A | N + M, t] = \Pr[t + R \geq A | M, t]$$

or since t is already known

$$P = \Pr[R \geq A - t | M, t].$$

We can calculate the above quantity, for example, if θ is assumed uniformly distributed a priori to be

$$P = \Pr[R \geq A - t]$$

$$= \begin{cases} \sum\limits_{r=A-t}^{M} \binom{t+r}{t}\binom{N+M-t-r}{N-t} \Big/ \binom{N+M+1}{N+1} & \text{if } A \geq t \\ 1 & \text{if } A \leq t \\ 0 & \text{if } N-t \\ & \qquad > N+M-A. \end{cases}$$

(10.2)

Now this is the probability that the goal will be reached by the end of the trial. If this is small enough, clearly there is not much point in continuing the trial.

If θ is assumed to have a beta prior density $p(\theta) \propto \theta^{a-1}(1 - \theta)^{b-1}$, then the summation on the r.h.s. of (10.2) is replaced by

$$\sum_{r=A-t}^{M} \binom{M}{r} \frac{\Gamma(r+t+a)\Gamma(N+M-r-t+b)\Gamma(N+b+a)}{\Gamma(N+M+a+b)\Gamma(t+a)\Gamma(N+b-t)}$$

$$\text{if } A \geq t.$$

If the trial consists of a standard and a new treatment and for $N_1 + M_1$ trials on the standard and $N_2 + M_2$ on the new treatment there may be various values for (Y_1, Y_2) that indicate a new treatment superior to the standard, where Y_1 and Y_2 are the final number of successes if the trial were carried to completion. Since for any given pair (N_1, N_2) we have $(Y_1 = t_1 + R_1, Y_2 = t_2 + R_2)$, we could calculate the predictive distribution for R_1 and R_2 given (N_1, t_1) and (N_2, t_2) given priors for θ_1 and θ_2. We then suppose that the criterion for the new agent to be better than the standard will require that $(Y_1, Y_2) \in B_S$, which could imply $(R_1, R_2) \in B_R$ say. We would then calculate the probability of $(R_1, R_2) \in B_R$ to aid in ascertaining the worth of continuing the trial to its conclusion. Of course a determination to continue the trial to its conclusion will usually depend on other factors as well.

Numerical illustration of Example 10.1. Candidates A and B are running for the office of Mayor. A poll of 80 voters is to be taken. The first 75 votes tallied were 38 to 37 in favor of candidate A. If the entire poll of 80 were taken, what is the predictive probability that candidate A will win the poll and also be the eventual winner? Assuming a uniform prior, the probability that A wins the poll is

$$\Pr\left[\sum_1^{75} x_i + \sum_{76}^{80} X_i \geq 41 \right] = 0.5119.$$

The probability that A will be the eventual winner of the election assuming the total number of voters is very large is

$$\lim_{M \to \infty} [Y/(N+M) > 0.5] = \Pr\left[\theta > 0.5 \middle| \sum_1^{75} x_i = 38 \right]$$

$$= \int_{0.5}^1 \frac{\theta^{38}(1-\theta)^{37}}{B(39,38)} d\theta = 0.5066.$$

Example 10.2. Suppose $X_1, \ldots, X_N, X_{N+1}, \ldots, X_{N+M}$ are independently and identically distributed as $N(\mu, 1)$. Suppose that we wish to test $H_0: \mu \leq a$ vs. $H_1: \mu > a$. Assume a prior for μ, $p(\mu)$ that leads to

$$p(\mu | x^{(N+M)}).$$

We then suppose that we will decide that $\mu \leq a$ if and only if

$$\Pr(\mu \leq a | x^{(N+M)}) \geq p$$

for some arbitrarily specified p.

Now for the sake of an example assume that $p(\mu) \propto$ const., so that the posterior distribution of μ is $N(\bar{x}_{N+M}, 1/(N+M))$. Then

$$\Pr(\mu \leq a | x^{(N+M)}) = \Phi\left(\frac{a - \bar{x}_{N+M}}{\sqrt{1/(N+M)}} \right) \geq p$$

or

$$\sqrt{N+M}\,(a - \bar{x}_{N+M}) \geq \Phi^{-1}(p)$$

or

$$(N+M)a - \Phi^{-1}(p)\sqrt{N+M} - N\bar{x}_N \geq M\bar{x}_M.$$

Now if we have already observed $X_1, \ldots, X_N$ then we can compute for fixed $\bar{x}_N$ and future $\bar{X}_M$

$$P = \Pr\left[\bar{X}_M \leq \frac{a(N+M)}{M} - \frac{(N+M)^{1/2}\Phi^{-1}(p)}{M} - \frac{N}{M}\bar{x}_N\right].$$

Since the predictive distribution of $\bar{X}_M$ is $N(\bar{x}_N, (1/M) + (1/N))$ then

$$P = \Phi\left\{\left(\frac{N}{M}\right)^{1/2}\left[(a - \bar{x}_N)(N+M)^{1/2} - \Phi^{-1}(p)\right]\right\}.$$

Clearly for fixed values a, N, M, and p, P increases monotonically with decreasing $\bar{x}_N$. Note also that, irrespective of p,

$$\lim_{M \to \infty} P = \Phi\left[N^{1/2}(a - \bar{x}_N)\right] = \Pr\left[\mu \leq a | x^{(N)}\right].$$

Example 10.3. Suppose now $X_1, \ldots, X_N, X_{N+1}, \ldots, X_{N+M}$ are independently and identically distributed as

$$f(x) = \theta\,e^{-\theta x}.$$

Assume that we wish to test H_0: $\theta \leq a$ versus H_1: $\theta > a$. We assume a prior density $p(\theta)$ for θ, and decide that $\theta \leq a$ if

$$\Pr(\theta \leq a | x^{(N+M)}) \geq p.$$

Suppose we assume that $p(\theta) \propto \theta^{-1}$ so that the posterior distribution of $2\theta(N+M)\bar{x}_{N+M}$ is a χ^2_{2N+2M} variate. Hence we require

that

$$\Pr(\theta \leq a \mid x^{(N+M)}) = F\left[2a(N+M)\bar{x}_{N+M}\right] \geq p$$

where $F(\cdot)$ is the distribution function of a χ^2_{2N+2M} variate.
Further

$$2a(N+M)\bar{x}_{N+M} \geq F^{-1}(p)$$

$$M\bar{x}_M \geq \frac{1}{2a}F^{-1}(p) - N\bar{x}_N.$$

Now if we stopped after the first N observations we can calculate the predictive probability

$$P = \Pr\left[M\bar{X}_M \geq \frac{1}{2a}F^{-1}(p) - N\bar{x}_N\right].$$

It is easy to show that the predictive distribution of

$$\bar{X}_M \sim \bar{x}_N Y$$

where Y is an F-variate with $2M$ and $2N$ degrees of freedom. Hence

$$P = \Pr\left(\frac{\bar{X}_M}{\bar{x}_N} \geq \frac{F^{-1}(p)}{2aM\bar{x}_N} - \frac{N}{M}\right)$$

$$= 1 - F_{2M,2N}\left(\frac{1}{2aM\bar{x}_N}F^{-1}(p) - \frac{N}{M}\right),$$

where $F_{2M,2N}(y)$ is the distribution function of Y. Further, for any $p \in (0, 1)$

$$\lim_{M \to \infty} P = \Pr[\theta \leq a \mid x^{(N)}]$$

the posterior probability after observing $x^{(N)}$.

Example 10.4. Suppose a random sample from a Poisson distribution is to be taken where

$$f(x) = \frac{e^{-\lambda}\lambda^x}{x!} \qquad x = 0, 1, \ldots.$$

For a test H_0: $\lambda > a$ versus H_1: $\lambda \leq a$, we decide that $\lambda \leq a$ if and only if

$$Pr(\lambda \leq a | x^{(N+M)}) \geq p.$$

If we assume that $p(\lambda) \propto \lambda^{-1}$, then for

$$y = \sum_{i=1}^{N+M} x_i = y_N + y_M \qquad y_N = \sum_1^N x_i, \qquad y_M = \sum_{N+1}^{N+M} x_i$$

$$p(\lambda | y) = \frac{\lambda^{y-1}(N+M)^y e^{-(N+M)\lambda}}{(y-1)!}, \qquad \lambda > 0, \qquad y > 0.$$

Hence $2(N+M)\lambda$ is a χ^2 variate with $2y$ degrees of freedom, and

$$Pr[\lambda \leq a | y] = F_{2y}[2(N+M)a] \geq p$$

with F_{2y} the distribution function of a χ^2 with $2y$ degrees of freedom. Further

$$2(N+M)a \geq F_{2y}^{-1}(p).$$

Then after observing y_N we need to calculate

$$P = Pr\left[F_{2y_N+2Y_M}^{-1}(p) \leq 2(N+M)a \right]$$

using the distribution of $Y_M = \sum_{N+1}^{N+M} X_i$ conditional on y_N.

Now the predictive probability function of Y_M given y_N is easily obtained as a negative binomial where

$$Pr[Y_M = t | y_N] = \binom{y_N + t - 1}{t} \left(\frac{M}{M+N} \right)^t \left(\frac{N}{M+N} \right)^{y_N}$$

for $t = 0, 1 \ldots.$

Further, the actual computation

$$P = \Pr\left[F^{-1}_{2y_N + 2Y_M}(p) \leq 2(N + M)a\right]$$

is more easily accomplished by finding the largest value of y which satisfies

$$F_{2y}[2(N + M)a] \geq p$$

say y_a. Then we calculate

$$P = \begin{cases} \displaystyle\sum_{t=0}^{y_a - y_N} \binom{y_N + t - 1}{t}\left(\frac{M}{M+N}\right)^t \left(\frac{N}{M+N}\right)^{y_N} & \text{if } y_a \geq y_N > 0 \\ 0 & \text{if } y_a < y_N. \end{cases}$$

Further it can also be shown that

$$\lim_{M \to \infty} P = F_{2y_N}(2Na) = \Pr\left[\lambda \leq a | y_N\right]$$

irrespective of p.

Numerical illustration of Examples 10.3 *and* 10.4. Requests for assistance by a towing service are assumed to follow a Poisson process with rate λ per minute. In a period of 30 minutes 20 requests are received. Suppose we wish to test, for $a = \frac{1}{2}$,

$$H_0: \lambda \geq \tfrac{1}{2} \text{ vs. } H_1: \lambda < \tfrac{1}{2}$$

and will decide in favor of H_1 if at the end of $N + M = 51$ minutes of observation

$$\Pr\left[\lambda < \tfrac{1}{2} | x^{(N+M)}\right] \geq 0.95.$$

After 30 minutes of observation we note that $2(N + M)a = 51$, so that

$$\Pr\left[\chi^2_{36} \leq 51\right] = 0.95$$

and $2y_a = 36$ or $y_a = 18 < 20 = y_{30}$, therefore $P = 0$.

On the other hand as before if it took 30 minutes to receive 20 requests for assistance and we want to test the above hypothesis by continuing observation until 14 additional requests are made, then $N = 20$, $\Sigma_1^N x_i, = 30$, $M = 14$, and

$$P = 1 - F_{28.40}\left[\frac{F^{-1}(0.95)}{21} - \frac{10}{7}\right] = 0.0016.$$

which is quantitatively larger than before even though we expect the extra 14 requests to take 21 minutes.

Example 10.5. Let X_i, $i = 1, \ldots, N + M$ be independently and identically distributed as $N(\mu, \sigma^2)$. For a test of $\sigma^2 < a$ versus $\sigma^2 \geq a$ we decide that $\sigma^2 \geq a$ if

$$\Pr(\sigma^2 \geq a | x^{(N+M)}) \geq p.$$

In particular, let

$$p(\mu, \sigma^2) \propto \frac{1}{\sigma^2}.$$

Then the posterior distribution of $\nu s_{N+M}^2 / \sigma^2$ is χ_ν^2 for $\nu = N + M - 1$, where

$$\nu s_{N+M}^2 = \sum_1^{N+M} (x_i - \bar{x}_{N+M})^2 \quad \text{and} \quad (N+M)\bar{x}_{N+M} = \sum_1^{N+M} x_i.$$

Hence

$$\Pr(\sigma^2 \geq a | x^{(N+M)}) = P\left(\frac{\nu s_{N+M}^2}{\sigma^2} \leq \frac{\nu s_{N+M}^2}{a}\right)$$

$$= F_\nu\left(\frac{\nu s_{N+M}^2}{a}\right)$$

where F_ν is the distribution function of a χ_ν^2 variate. Now for

$$F_\nu\left(\frac{\nu s_{N+M}^2}{a}\right) \geq p$$

$$\frac{\nu s_{N+M}^2}{a} \geq F_\nu^{-1}(p).$$

Since

$$\nu s_{N+M}^2 = (N-1)s_N^2 + (M-1)s_M^2 + \frac{NM}{N+M}(\bar{x}_M - \bar{x}_N)^2,$$

where

$$(N-1)s_N^2 = \sum_1^N (x_i - \bar{x}_N)^2, \qquad N\bar{x}_N = \sum_1^N x_i,$$

$$(M-1)s_M^2 = \sum_{N+1}^{N+M} (x_i - \bar{x}_M)^2$$

and

$$M\bar{x}_M = \sum_{N+1}^{N+M} x_i$$

we can, after observing $x^{(N)}$, calculate

$$P = \Pr\left[(N-1)s_N^2 + Y + \frac{NM}{N+M}(\bar{X}_M - \bar{x}_N)^2 \geq aF_\nu^{-1}(p)\right]$$

$$= \Pr\left[Y + \frac{NM}{N+M}(\bar{X}_M - \bar{x}_N)^2 \geq aF_\nu^{-1}(p) - (N-1)s_N^2\right]$$

where Y is the random variable representing the unobserved $(M-1)s_M^2$. Hence in order to evaluate the above we must calculate the joint predictive distribution of Y and $\bar{X}_M$ given $x^{(N)}$. First we note for $N > 1$

$$p(\mu, \sigma^2 | x^{(N)}) = \frac{z^{(N-1)/2} e^{-z/2\sigma^2}\sqrt{N} e^{-(N/2\sigma^2)(\bar{x}_N - \mu)^2}}{2^{(N-1)/2}\sigma^{N+1}\Gamma[(N-1)/2]\sigma\sqrt{2\pi}}$$

where $z = (N-1)s_N^2$. Setting $\bar{X}_M = X$, the predictive density of X and Y is

$$f(x, y | x^{(N)}) = \int f(x, y | \mu, \sigma^2) p(\mu, \sigma^2 | x^{(N)}) \, d\mu \, d\sigma^2$$

where for $M > 1$

$$f(x, y|\mu, \sigma) = \frac{\sqrt{M}\, e^{-(M/2\sigma^2)(x-\mu)^2} y^{[(M-1)/2]-1} e^{-z/2\sigma^2}}{\sigma\sqrt{2\pi}\, 2^{(M-1)/2}\Gamma[(M-1)/2]\sigma^{M-1}}.$$

Since

$$M(x-\mu)^2 + N(\bar{x}_N - \mu)^2$$
$$= (N+M)(\mu - w)^2 + \frac{NM}{N+M}(x - \bar{x}_N)^2$$

where $(N+M)w = N\bar{x}_N + Mx$, integration with respect to μ and σ yield for $M > 1$

$$f(x, y|x^{(N)})$$
$$= \frac{\sqrt{MN}\,\Gamma[(M+N-1)/2]\, z^{(N-1)/2} y^{(M-3)/2}}{\sqrt{M+N}\,\Gamma(1/2)\Gamma[(N-1)/2]\,\overline{\Gamma[(M-1)/2]}}$$
$$\times \left[z + y + \frac{NM}{N+M}(x - \bar{x}_N)^2 \right]^{-(M+N-1)/2}. \qquad (10.3)$$

Now we need to calculate

$$P = \Pr\left\{ \frac{Y + [NM/(N+M)](X - \bar{x}_N)^2}{z} \geq \frac{aF_\nu^{-1}(p)}{z} - 1 \right\}$$

or

$$P = \Pr\left\{ \frac{(N-1)\left[Y + [NM/(N+M)](X - \bar{x}_N)^2\right]}{Mz} \right.$$
$$\left. \geq \frac{a(N-1)F_\nu^{-1}(p)}{Mz} - \frac{N-1}{M} \right\}.$$

From the joint predictive density of X and Y we can easily show that

$$\frac{(N-1)\left\{Y + [NM/(N+M)](X - \bar{x}_N)^2\right\}}{Mz}$$

is an F-variate with M and N-1 degrees of freedom. Hence

$$P = 1 - F_{M,N-1}\left[\frac{aF_\nu^{-1}(p)}{Ms_N^2} - \frac{N-1}{M}\right].$$

It can easily be shown that the above holds for $M = 1$ as well. Again for any $p \in (0,1)$

$$\lim_{M \to \infty} P = \Pr[\sigma^2 \geq a|x^{(N)}].$$

Example 10.6. Under the same circumstances as in the previous example suppose we wish to test $\mu \leq a$ versus $\mu > a$. Now the posterior density of

$$\left[(\mu - \bar{x}_{N+M})\sqrt{N+M}\right]\Big/ s_{N+M}$$

is t with $\nu = N + M - 1$ degrees of freedom. Hence we decide on $\mu \leq a$ if

$$\Pr[\mu \leq a|x^{(N+M)}] = S_\nu\left[\frac{(a - \bar{x}_{N+M})\sqrt{N+M}}{s_{N+M}}\right] \geq p,$$

where $S_\nu(\cdot)$ is the distribution function of t with ν degrees of freedom. Taking the inverse function we obtain

$$\frac{(a - \bar{x}_{N+M})\sqrt{N+M}}{s_{N+M}} \geq S_\nu^{-1}(p).$$

After observing $x^{(N)}$ we need to calculate

$$P =$$

$$\Pr\left\{ \frac{\left(a - \dfrac{N\bar{x}_N}{N+M} - \dfrac{MX}{N+M}\right)(N+M)^{1/2}(N+M-1)^{1/2}}{\left(z + Y + \dfrac{NM}{N+M}(X - \bar{x}_N)^2\right)^{1/2}} \geq S_\nu^{-1}(p) \right\}.$$

where the random variable on the left within the bracket is a

for fixed $\bar{y}_N - \bar{x}_N$ and random $(\bar{Y}_M - \bar{X}_M)$, where $\Phi^{-1}(p)$ is the inverse function of Φ.

Hence

$$P = \Pr\left[M\left(\bar{Y}_M - \bar{X}_M\right) \geq \Phi^{-1}(p)\sigma\sqrt{2(N+M)} \right.$$
$$\left. - N\left(\bar{y}_N - \bar{x}_N\right) + (N+M)\delta \right].$$

Now the predictive distribution of

$$\bar{Y}_M - \bar{X}_M \sim N\left[\bar{y}_N - \bar{x}_N, 2\sigma^2\left(\frac{1}{M} + \frac{1}{N}\right) \right]$$

so that

$$P = \Phi\left\{ \left(\frac{N}{M}\right)^{1/2}\left[\frac{(\bar{y}_N - \bar{x}_N - \delta)(N+M)^{1/2}}{\sigma\sqrt{2}} - \Phi^{-1}(p) \right] \right\}.$$

$$(10.4)$$

The above can now be computed for any M to help determine whether it is worthwhile to continue the experiment. Note also that as M increases

$$\lim_{M \to \infty} P = \Phi\left(\frac{(\bar{y}_N - \bar{x}_N - \delta)\sqrt{N}}{\sigma\sqrt{2}} \right) = \Pr\left[\theta > \delta \,|\, \bar{x}_N, \bar{y}_N \right]$$

independent of $p \in (0,1)$. Of course this is the probability of H_0 after N observations on both groups are in hand essentially with probability 1. This is a very simple and easily computable example. If the variance were unknown the exact calculation would be considerably more difficult. But for a reasonable sample size N the following substitutions should be adquate: The pooled sample variance

$$s^2 = \frac{1}{2(N-1)}\left[\sum_{i=1}^{N}(x_i - \bar{x})^2 + \sum_{i=1}^{N}(y_i - \bar{y})^2 \right]$$

for σ^2; $S^{-1}_{2N+2M-2}(p)$ for $\Phi^{-1}(p)$ and $S_{2N-2}(\cdot)$ for $\Phi(\cdot)$ in (10.4), where $S_\nu(\cdot)$ represents the distribution function of a student "t" with ν degrees of freedom. The extension to samples of unequal nterim and final size i.e. N_1 and N_2 and M_1 and M_2 can easily be made.

Numerical illustration of Example 10.7. In a hospital unit that contains 60 beds a random sample of 60 women between 50 and 60 years of age was chosen to test a new soporific drug versus a placebo. The drug Y was given to 30 of the women chosen at random from the 60 and the rest received the placebo X. Average sleep time per day over a week was recorded for each woman. Sleep time in hours was assumed to be normally distributed with the same variance for each group. The data for the two agents are

$$\bar{x}_{30} = 6.4, \qquad \bar{y}_{30} = 7.8, \qquad s^2 = 1.8$$

and the hypothesis to be tested was

$$H_0: \theta \geq 1 \text{ vs. } H_1: \theta < 1.$$

We note that after observing the 60 women

$$\Pr\left[\theta \geq 1 | \bar{x}_{30}, \bar{y}_{30}, s^2\right] = 0.876.$$

We assume that a regulatory agency requires that before a drug is deemed potentially effective the posterior probability of $\theta \geq 1$ must be at least 0.9 and that the drug has been tested on at least 50 individuals both for the placebo and the drug. The predictive probability that the new drug will be deemed effective if another 20 or 30 women were placed on each of the two agents is calculated to be 0.60 for $M = 20$ and 0.64 for $M = 30$. For this example we have also graphed P for various values of M and $\gamma = 0.8, 0.9,$ and 0.95 (Fig. 10.1).

Example 10.3 (*continued*). Under the previous assumptions we take up the problem of censored observations where the censoring does not depend on θ. After an interim sample of size N at some given time $x_1, \ldots, x_d$ are fully observed and the rest $x_{d+1}, \ldots, x_N$ are censored in the sense of being lost to follow-up at or prior to the interim assessment. Now $2\theta(N + M)\bar{x}_{N+M}$ is distributed a posteriori as a χ^2_{2d+2k} variate where $\sum_1^{N+M} x_i = (N + M)\bar{x}_{N+M}$

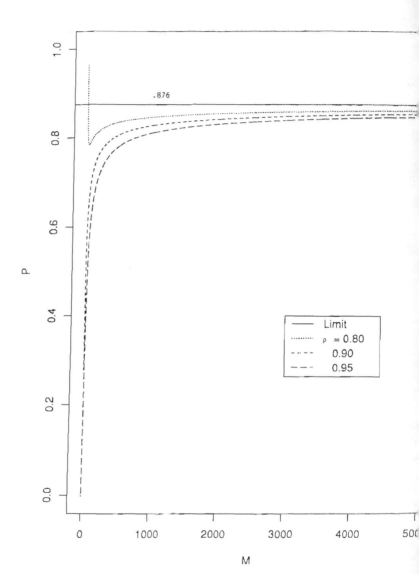

Figure 10.1 *Values of P for varying p and M*

and k is the number of uncensored observations out of M future observations. To decide that $\theta \le a$, we require the occurrence of the event

$$A = \left\{ \Pr[\theta \le a | x^{(N+M)}] = F[2a(N+M)\bar{x}_{N+M}] \ge p \right\}$$

where $F(\cdot)$ is the distribution function of a χ^2_{2d+2k} variate. This implies that, for the future values $X_{N+1}, \ldots, X_{N+M}$ we need to calculate the probability of A, which is for fixed k

$$P(A|k) = \Pr\left[M\bar{X}_M \ge \frac{1}{2a}F^{-1}(p) - N\bar{x}_N \right].$$

Now the predictive distribution of $\bar{X}_M$ given $\bar{x}_N$ is such that

$$\frac{M\bar{X}_M}{2k} \sim \frac{N\bar{x}_N}{2d}Y$$

where Y is an F-variate with $2k$ and $2d$ degrees of freedom. Therefore

$$P(A|k) = 1 - F_{2k,2d}\left(\frac{dF^{-1}(p)}{2akN\bar{x}_N} - \frac{d}{k} \right)$$

if k were known. However, k, the number of future uncensored values, can be taken into account in a variety of ways. We can simply assume that the future experiment will not be concluded until $k = M$, or some other value k_0. An alternative is to let the number of future uncensored values be a random variable K with $\Pr[K = k]$, $k = 0, 1, \ldots, M$ depending on the censoring mechanism assumed for the future experiment of size M. For example we can set $\Pr(K = k_0) = 1$ indicating a future experiment that would terminate as soon as k_0 values had failed. A more likely scenario is that the future experiment would be censored after a given value x_0 had been achieved, especially if we were dealing with future survival times, thus requiring the calculation of $\Pr[K = k | x_0]$. In such a case we then would calculate

$$P = 1 - \sum_{k=0}^{M} \Pr[K = k | x_0] I_{k, x_0}$$

where

I_{k, x_0}

$$= \begin{cases} 0 & \text{if } (M - k)x_0 \geq \dfrac{1}{2a} F^{-1}_{2(k+d)}(p) - N\bar{x}_N \\[3mm] 1 & \text{if } k = 0 \text{ and } Mx_0 < \dfrac{1}{2a} F^{-1}_{2d}(p) - N\bar{x}_N \\[3mm] F_{2k, 2d}\left[\dfrac{dF^{-1}(p)}{2akN\bar{x}_N} - \dfrac{d}{k} \right] & \text{if } (M - k)x_0 < \dfrac{1}{2a} F^{-1}_{2(k+d)}(p) - N\bar{x}_N \end{cases}$$

In this case

$$\Pr[K = k | x_0, \theta] = \binom{M}{k} (e^{-\theta x_0})^{M-k} (1 - e^{-\theta x_0})^k$$

$$\Pr[K = k | x_0] = \int p(\theta | x^{(N)}) \Pr[K = k | x_0, \theta]\, d\theta$$

$$= \binom{M}{k} (N\bar{x}_N)^d \sum_{j-0}^{k} (-1)^j \binom{k}{j}$$

$$\times [N\bar{x}_N + (M - j)x_0]^{-d}.$$

The situation can be more complex than this since at the interim analysis there may be censored observations that are not lost to follow-up but can continue to be observed. This requires that the initial censored observations be partitioned such that $x_{d+1}, \ldots, x_{d+l}$ are lost to follow-up prior to the interim analysis and $x_{d+l+1}, \ldots, x_N$ are censored at the time of the interim analysis. Although the latter were all censored at the same time their values may be different if they were treated or put on test at different times. We now assume that M further observations will be taken and that among these k, i.e., $x_{N+1}, \ldots, x_{N+k}$ will be fully observed at the end of the experiment while the rest $x_{N+k+1}, \ldots, x_{N+M}$ will be censored, i.e., some lost to follow-up and others still surviving. Now out of the $N - d - l$ observations, $x_{d+l+1}, \ldots, x_N$ that have survived up to the interim analysis t will have been fully observed (failed) and $N - d - l - t$ will still have survived by the end of the entire experiment. Hence it is easily shown that the posterior distribution of $2\theta \sum_1^{N+M} x_i$ is $\chi^2_{2d+2k+2t}$.

Now at the end of the experiment we will decide for H_0 if the event

$$A = \left\{ \Pr[\theta \leq a | x^{(N+M)}] \geq p \right\}$$

holds or if

$$F\left(2a \sum_{1}^{N+M} x_i \right) \geq p,$$

or

$$\sum_{1}^{N+M} x_i \geq \frac{1}{2a} F^{-1}(p) \tag{10.5}$$

where F is the distribution function of a $\chi^2_{2d+2k+2l}$ variate and $F^{-1}(p)$ is the inverse function of F.

For $i = d + l + 1, \ldots, N$, let $X_i - x_i = Y_i$, where x_i is the censored value, i.e., the value at the time of the interim analysis. Because of the memoryless property of the exponential distribution Y_i has the original exponential distribution. Hence from (10.5) we obtain for future random variables

$$\sum_{i=N+1}^{N+M} X_i + \sum_{i=d+l+1}^{N} Y_i \geq \frac{1}{2a} F^{-1}(p) - \sum_{1}^{N} x_i. \tag{10.6}$$

Since the predictive distribution of

$$\frac{2d\left(\sum_{i=N+1}^{N+M} X_i + \sum_{i=d+l+1}^{N} Y_i \right)}{2(k+l)\sum_{1}^{N} x_i} \tag{10.7}$$

is easily found to be an F-variate with $2k + 2l$ and $2d$ degrees of freedom, we calculate, for $k + l \geq 1$, the probability P that continuing the experiment will lead to acceptance of H_0. The result is

$$\Pr\left[\sum_{i=N+1}^{N+M} X_i + \sum_{i=d+l+1}^{N} Y_i \geq \frac{1}{2a} F^{-1}(p) - \sum_{1}^{N} x_i \right]$$

$$= 1 - F_{2(k+l),2d}\left[\frac{dF^{-1}(p)}{2a(k+l)\sum_{1}^{N} X_i} - \frac{d}{k+l} \right] \tag{10.8}$$

where $F_{2(k+t),2d}(\cdot)$ is the distribution function of the F-variate. The result above presupposes that the future experiment will terminate when one achieves exactly k and t failures. However, the more likely scenario is that the trial will terminate at a given time x_0, say. Now the future number of uncensored values k and t are random variables depending on the time x_0. However (10.8) depends only on the sum $k + t = r$, i.e., it is the conditional predictive probability of A given r. Let K and T stand for the random variables, which are observed as k and t. Then for $R = K + T$ and $J = M + N - d - l$

$$\Pr[R = r|x_0, \theta] = \binom{J}{r}(e^{-\theta x_0})^{J-r}(1 - e^{-\theta x_0})^r.$$

Then

$$\Pr[R = r|x_0] = \int \Pr[R = r|x_0, \theta] p(\theta|x^{(N)}) \, d\theta$$

$$= \binom{J}{r}\left(\sum_1^N x_i\right)^d \sum_{u=0}^r (-1)^u \binom{r}{u}\left[\sum_1^N x_i + (J-u)x_0\right]^{-d}$$

$$(10.9)$$

Hence we calculate the unconditional predictive probability of A denoted as

$$P_{x_0, M} = \sum_{r=0}^J P(A|r, x_0)\Pr(R = r|x_0)$$

where

$$P(A|r, x_0) = 1 - I_{r, x_0}$$

$$I_{0, x_1} = \begin{cases} 1 & \text{if } Jx_0 < \dfrac{1}{2a}F_{2d}^{-1}(p) - \sum_1^N x_i \\ 0 & \text{otherwise} \end{cases}$$

and, for $r \geq 1$,

$$I_{r, r_0}$$

$$= \begin{cases} F_{2r, 2d} \left(\dfrac{dF_{2(r+d)}^{-1}(p)}{2ar\sum_1^N x_i} - \dfrac{d}{r} \right) & \text{if } (J - r)x_0 \\ & \qquad < \dfrac{1}{2a} F_{2(r+d)}^{-1}(p) - \sum_1^N x_i \\ 0 & \text{otherwise,} \end{cases}$$

and F_ν^{-1} is the inverse function of the χ_ν^2 distribution. If a conjugate prior on θ, namely

$$p(\theta) \propto \theta^{\delta - 1} e^{-\gamma \theta}$$

were used, then we would merely substitute $\delta + d$ for d, $\sum x_i + \gamma$ for $\sum x_i$ and $F^{-1}(p)$ is now the inverse function of a $\chi_{2(d+\delta+r)}^2$ distribution in all the previous results.

Numerical illustration of Example 10.3 *(continued)*. Suppose in the personnel carrier illustration of Example 3.7 it was learned that the vehicle with 508 miles was in an accident irrelevant to testing and thus was lost to followup. Also we assume that the two vehicles with 2358 and 2880 miles, respectively, did not actually fail but were censored at those values. The acceptance of this type of vehicle depends, say, on testing at least 25 vehicles and asserting, for $1/a = 950$,

$$\Pr\left[\frac{1}{\theta} > 950 | x^{(n)} \right] \geq 0.9. \tag{10.10}$$

Now at the interim point $N = 19$, $d = 2$, $l = 1$, $\sum_{i=1}^{19} x_i = 18{,}947$ and assuming $\gamma = 0$,

$$\Pr\left[\frac{1}{\theta} > 950 | x^{(19)} \right] = 0.841.$$

We now compute the chance that (10.10) is true if we put M additional vehicles on test and terminated testing future vehicles when they either fail or attain $x_0 = 3000$ miles, whichever comes first. The results obtained are $0.502, 0.565, 0.609$ for $M = 10$, 20, and 30, respectively. In Figure 10.2, there are graphs of P for $M = 10$, 20, 30, and ∞ for varying values of $1/a$ up to 2500.

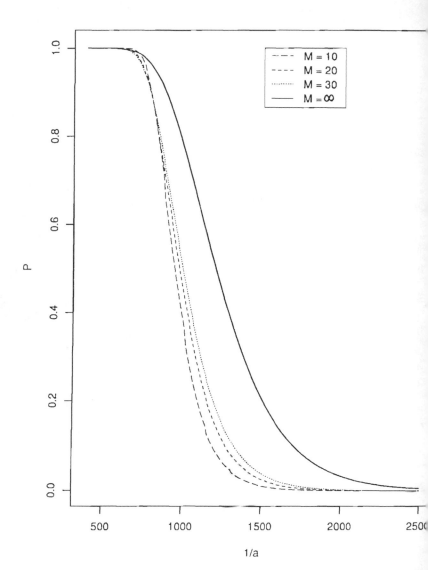

Figure 10.2

Example 10.8. Suppose X_i $i = 1, \ldots, n$ are independently $N(\mu, \Sigma)$ where Σ is known. Let

$$H_0: \mu \in Q_0 \text{ vs. } H_1: \mu \notin Q_0$$
$$Q_0 = \{\mu: (\mu - \mu_0)'\Sigma^{-1}(\mu - \mu_0) \leq t_0\}$$

for a given $t_0 > 0$. Then we will decide for H_0 if

$$\Pr[\mu \in Q_0 | x^{(n)}] \geq p$$

where say $p(\mu) \propto$ const. so that

$$\mu \sim N(\bar{x}, (1/n)\Sigma).$$

Now

$$V = n(\mu - \mu_0)'\Sigma^{-1}(\mu - \mu_0) \sim \chi_p^2(\lambda)$$

a noncentral chi-squared variate with noncentrality parameter

$$\lambda = n(\bar{x} - \mu_0)'\Sigma^{-1}(\bar{x} - \mu_0)$$

for $\bar{x} = n^{-1}\sum_{i=1}^n x_i$. Hence, we accept H_0 if

$$\Pr[V \leq nt_0 | x^{(n)}] \geq p.$$

Assume we want to conduct an interim analysis at N to provide guidance on whether to continue the experiment. Hence setting $n = N + M$ we need to calculate

$$P = \Pr\left[\Pr(V \leq (N + M)t_0 | x^{(N)}, X_{(M)}) \geq p\right].$$

Now

$$\bar{x} = \frac{M\bar{X}_M + N\bar{x}_N}{N + M}$$

and setting $\bar{X}_M = Y$, $\bar{x}_N = x$, and $G(\cdot|\lambda)$ as the distribution function of the noncentral $\chi_p^2(\lambda)$

$$P = \Pr\{G[(N + M)t_0|\lambda] \geq p\}. \tag{10.11}$$

Now

$$Y|x \sim N\left[x, \Sigma\left(\frac{1}{M} + \frac{1}{N}\right)\right]$$

and hence

$$\lambda = \lambda(Y, x) \sim \frac{M}{N}\chi^2_M(\rho)$$

where

$$\rho = (x - \mu_0)'\Sigma^{-1}(x - \mu_0).$$

Further, it can be shown that $1 - G[(N + M)t_0|\lambda]$ is increasing in λ. Therefore the calculation of P involves calculating the minimum value of λ, say λ_0, such that $1 - G[(N + M)t_0|\lambda] \leq p$ and then

$$P = \Pr(\lambda > \lambda_0)$$

from the noncentral chi-squared distribution of λ.

Further, when Σ is unknown and the usual $p(\mu, \Sigma^{-1}) \alpha$ $|\Sigma|^{(p+1)/2}$ is used, we have the more complex situation where V is distributed as a weighted sum of an infinite number of χ^2 variates with negative binomial weights (Geisser, 1967). This is similar to a noncentral chi-square variate that has Poisson weights. However, the monotonicity property in ρ still holds. Further, if we estimate the sample covariance matrix S_{N+M} by the now fixed S_N, as in the univariate case, it can be shown that ρ is distributed as a weighted sum of an infinite number of beta variates with negative binomial weights (Geisser and Johnson, 1993a). Hence the same type of approximate analysis can be carried out with these two distributions. This can be extended to a comparison of two populations and also be useful in determining whether further sampling will improve error rates in problems of classification (Geisser and Johnson, 1993b).

Perhaps an even more interesting H_0 is where the region is such that each component of μ, say $\mu_i \geq a_i$. In this situation, simulation procedures are necessary for the computation of the requisite integrals.

10.4 Applications to predictive hypotheses

The same idea can be used for a future observable. Suppose Z is a future observable and we wish to decide whether $\Pr(Z \in R) \geq p$, where R is some defined region. Of course, as before, a Bayesian can stop any time during sampling, say after N observations and calculate $\Pr(Z \in R | x^{(N)})$. However, it is of some interest to make calculations of the sort

$$\Pr\left[\Pr(Z \in R | x^{(N+M)}) \geq p\right] = P$$

after observing $x^{(N)}$.

We shall illustrate this with two cases. Assume $X_i \sim N(\mu, 1)$ $i = 1, \ldots, N + M$. The issue here could be to determine whether $\Pr(Z > a) \geq p$. A way of considering this is to determine whether

$$\Pr(Z > a | x^{(N+M)}) = \Phi\left(\frac{(a - \bar{x}_{N+M})(N+M)^{1/2}}{(N+M+1)^{1/2}}\right) \geq p.$$

Hence after N observations are in hand we can calculate

$$P = \Pr\left[\frac{(a - X)(N+M)^{1/2}}{(N+M+1)^{1/2}} \geq \Phi^{-1}(p) | x^{(N)}\right],$$

where the random variable

$$X = (N+M)^{-1}\left(N\bar{x}_N + M\bar{X}_M\right),$$

for fixed $\bar{x}_N$ and random $\bar{X}_M$. This is easily obtained as

$$P = \Phi\left\{\left(\frac{N}{M}\right)^{1/2}\left[(N+M)^{1/2}(a\bar{x}_N) - (N+M+1)^{1/2}\Phi^{-1}(p)\right]\right\}.$$

Again if P is large enough to suit one's purpose it may be worthwhile to continue sampling. Note also that if one sampled without end, then

$$\lim_{M \to \infty} P = \Phi\left[\sqrt{N}(a - \bar{x}_N) - \sqrt{N}\,\Phi^{-1}(p)\right].$$

Now this calculation depends on p, which was not the case for the parametric hypothesis on μ. Here if $p \to 1$, $P \to 0$, and if $p \to 0$, $P \to 1$ as expected. In the even odds case $p = 1/2$, P will be less than or greater than $1/2$ as $a - \bar{x}$ is positive or negative.

The case σ^2 is unknown will follow as well using the noninformative prior as before. Here we obtain

$$\Pr[Z > a|x^{(N+M)}] = S_\nu \left(\frac{(a - \bar{x}_{N+M})\sqrt{N+M}}{s_{N+M}\sqrt{N+M+1}} \right) \geq p$$

and now

$$P = \Pr\left[(a - X)\sqrt{N+M} \geq S_\nu^{-1}(p) s_{N+M}\sqrt{N+M+1}|x^{(N)}\right].$$

Using the approximation as in the parametric case we obtain

$$P \doteq S_{N-1}\left\{ \left(\frac{N}{M}\right)^{1/2}\left[\frac{(N+M)^{1/2}(a - \bar{x}_N)}{s_N} \right. \right.$$
$$\left. \left. - S_\nu^{-1}(p)\sqrt{N+M+1} \right] \right\}.$$

Further

$$\lim_{M \to \infty} P \doteq S_{N-1}\left(\frac{\sqrt{N}(a - \bar{x}_N)}{s_N} - \sqrt{N}\,\Phi^{-1}(p) \right).$$

For a sample of size $N + M$ from the exponential distribution the predictive distribution of Z is

$$\Pr[Z \leq a|x^{(N+M)}] = G\left(\frac{a}{\bar{x}_{N+M}} \right)$$

where $G(\cdot)$ is the distribution function of an F variate with 2 and

2($N + M$) degrees of freedom. Now we calculate

$$P = \Pr\left[G\left(\frac{a(N+M)}{N\bar{x}_N + M\bar{X}_M} \right) \geq p \right]$$

$$= \Pr\left[\frac{a(N+M)}{N\bar{x}_N + M\bar{X}_M} \geq G^{-1}(p) \right]$$

$$= \Pr\left[\frac{\bar{X}_M}{\bar{x}_N} \leq \frac{a(N+M)}{M\bar{x}_N G^{-1}(p)} - \frac{N}{M} \right]$$

$$= H\left[\frac{a(N+M)}{M\bar{x}_N G^{-1}(p)} - \frac{N}{M} \right]$$

where $H(\cdot)$ is the distribution function of an F variate with $2M$ and $2N$ degrees of freedom. Further

$$\lim_{M \to \infty} P = H_\infty\left(\frac{a}{\bar{x}_N G_\infty^{-1}(p)} \right)$$

where H_∞ represents the distribution function of $2N$ times the reciprocal of a χ^2_{2N} variate and G_∞ the distribution function of $(1/2)\chi^2_2$ random variate.

References

Bellman, R. (1957). *Dynamic Programming*. Princeton, NJ: Princeton University Press.

Geisser, S. (1967). Estimation associated with linear discriminants. *Annals of Mathematical Statistics* **38**, 807–817.

Geisser, S. (1992b). On the curtailment of sampling. *The Canadian Journal of Statistics* **20**(3), 297–309.

Geisser, S., and Johnson, W. O. (1993a). Interim analysis for normally distributed observables. *Proceedings of the International Symposium on Multivariate Analysis and its Applications* (to appear).

Geisser, S., and Johnson, W. O. (1993b). Sample size considerations for multivariate normal classification, (to appear).

Lindley, D. V., and Barnett, B. N. (1965). Sequential sampling: Two decision problems with linear losses for binomial and normal variables. *Biometrika* **52**, 507–532.

Bibliography

Aitchison, J. (1964). Bayesian tolerance regions. *Journal of the Royal Statistical Society B* **26**, 161–175.

Aitchison, J. (1970). Statistical problems of treatment allocation. *Journal of the Royal Statistical Society A* **133**, 206–239.

Aitchison, J. (1975). Goodness of fit prediction. *Biometrika* **62**(3), 547–554.

Aitchison, J., and Dunsmore, I. R. (1975). *Statistical Prediction Analysis*. Cambridge: Cambridge University Press.

Akaike, H. (1973). Information theory and an extension of the maximum likelihood principle. In B. N. Petrov and F. Czaki (eds.). *Proceedings of the 2nd International Symposium on Information Theory*, Budapest: Akademiai Kiado, 267–281.

Akaike, H. (1978). A new look at the Bayes procedure. *Biometrika* **65**, 53–59.

Andrews, D. F., and Hertzberg, A. M. (1985). *Data: A Collection of Problems from Many Fields for the Student and Research Worker*. Berlin: Springer-Verlag.

Andrews, D. F., and Pregibon, D. (1978). Finding outliers that matter. *Journal of the Royal Statistical Society B* **40**, 85–93.

Anderson, T. W. (1971). *The Statistical Analysis of Time Series*. New York: Wiley.

Atkinson, A. C. (1980). A note on the generalized information criterion for choice of a model. *Biometrika* **67**(2), 413–418.

Barnard, G. A. (1954). Sampling inspection and statistical decisions. *Journal of the Royal Statistical Society B* **16**, 151–174.

Barnett, V., and Lewis, T. (1978). *Outliers in Statistical Data*. New York: Wiley.

Bayes, T. (1763). An essay towards solving a problem in the doctrine of chances. *Philosophical Transactions of the Royal Society of London* **53**, 370–418.

Bellman, R. (1957). *Dynamic Programming*. Princeton, NJ: Princeton University Press.

Bernardo, J. M. (1979). Reference posterior distributions for Bayesian inference (with discussion). *Journal of the Royal Statistical Society B* **41**, 113-147.

Berliner, L. M., and Hill, B. M. (1988). Bayesian nonparametric survival analysis. *Journal of the American Statistical Association* **83**, 772-784.

Box, G. E. P. (1980). Sampling and Bayes' inference in scientific modelling and robustness. *Journal of the Royal Statistical Society B* **143**, 383-430.

Box, G. E. P., and Jenkins, M. (1970). *Time Series Analysis Forecasting and Control*. San Francisco: Holden-Day.

Box, G. E. P., and Tiao, G. C. (1973). *Bayesian Inference in Statistical Analysis*. Reading, MA: Addison-Wesley.

Brown, P. J. (1982). Multivariate calibration. *Journal of the Royal Statistical Society B* **44**, 287-321.

Burbea, J., and Rao, C. R. (1982). Entropy differential metric, distance and divergence measures in probability spaces: A unified approach. *Journal of Multivariate Analysis* **12**, 575-596.

Burkhardt, U., Mertens, Th., and Eggers, H. J. (1987). Comparison of two commercially available anti-HIV ELISAs: Abbott HTLV III EIAc and DuPont HTLV III-ELISAs. *Journal of Medical Virology* **23**, 217-224.

Butler, R. (1986). Predictive likelihood inference with applications. *Journal of the Royal Statistical Society B* **48**, 1-38.

Butler, R., and Rothman, E. D. (1980). Predictive intervals based on reuse of the sample. *Journal of the American Statistical Association* **75**(372), 881-889.

Cain, M., and Owen, R. J. (1990), Regression levels for Bayesian predictive response. *Journal of the American Statistical Association* **85**(409), 228-231.

Casella, G., and George, E. I. (1992). Explaining the Gibbs sampler. *The American Statistician* **46**(3), 167-174.

Chen, C. N. (1990). New diagnostic measures in the linear model. Unpublished doctoral dissertation. University of Minnesota.

Cook, R. D., and Weisberg, S. (1982). *Residuals and Influence in Regression*. New York: Chapman and Hall.

Cornfield, J. (1969). The Bayesian outlook and its applications. *Biometrics* **25**(4), 617-657.

Dawid, A. P. (1984). Statistical theory, the prequential approach. *Journal of the Royal Statistical Society A* **147**(2), 278-292.

de Finetti, B. (1937). Le Prevision: ses lois logiques, ses sources subjectives. *Ann. Inst. Poincare*, tome VIII, fasc. 1, 1-68. Reprinted in *Studies in Subjective Probability*. Melbourne, FL: Krieger, 1980 (English translation).

de Finetti, B. (1974). *Theory of Probability*. New York: Wiley (first published in 1970 under title of Teoria Delle Probabilita).

254   BIBLIOGRAPHY

De Groot, M. H. (1970). *Optimal Statistical Decisions.* New York: McGraw-Hill.

Dempster, A. P., and Gasko-Green, M. (1981). New tools for residual analysis. *Annals of Statistics* **9**, 945–959.

Devroye, L. (1987). *A Course in Density Estimation,* Boston: Birkhauser.

Dixon, W. J. (1950). Analysis of extreme values. *Annals of Mathematical Statistics* **21**, 488–506.

Dixon, W. J. (1951). Ratios involving extreme values. *Annals of Mathematical Statistics* **22**, 68–78.

Dunsmore, I. R. (1968). A Bayesian approach to calibration. *Journal of the Royal Statistical Society B* **30**, 396–405.

Dunsmore, I. R. (1969). Regulation and optimization. *Journal of the Royal Statistical Society B* **31**, 60–70.

Enis, P., and Geisser, S. (1974). Optimal predictive linear discrimination. *Annals of Statistics* **2**(2), 403–410.

Fisher, R. A. (1956). *Statistical Methods and Scientific Inference.* Edinburgh: Oliver and Boyd.

Freedman, D. A., and Purves, R. A. (1969). Bayes' method for bookies. *Annals of Mathematical Statistics* **40**, 1177–1186.

Gastwirth, J. L. (1987). The statistical precision of medical screening tests. *Statistical Science* **2**, 213–238.

Geisser, S. (1956). A note on the normal distribution. *Annals of Mathematical Statistics,* **27**, 858–859.

Geisser, S. (1964). Posterior odds for multivariate normal classification. *Journal of the Royal Statistical Society B* **1**, 69–76.

Geisser, S. (1965). Bayesian estimation in multivariate analysis. *Annals of Mathematical Statistics* **56**, 150–159.

Geisser, S. (1966). Predictive discrimination. In *Multivariate Analysis,* P. Krishnaiah (ed.). New York: Academic Press, 149–163.

Geisser, S. (1967). Estimation associated with linear discriminants. *Annals of Mathematical Statistics* **38**, 807–817.

Geisser, S. (1970). Bayesian analysis of growth curves. *Sankhya A* **32**, 53–64.

Geisser, S. (1971). The inferential use of predictive distributions. In *Foundations of Statistical Inference,* V. P. Godambe and D. A. Sprott (eds.). Toronto: Holt, Rinehart & Winston, 456–469.

Geisser, S. (1974). A predictive approach to the random effect model. *Biometrika* **61**, 101–107.

Geisser, S. (1975). The predictive sample reuse method with applications. *Journal of the American Statistical Association* **70**, 320–328.

Geisser, S. (1979). "Discussion" of Bernardo, "Reference posterior distributions for Bayesian inference." *Journal of the Royal Statistical Society B* **41**, 136–137.

Geisser, S. (1980a). Growth curve analysis. In *Handbook of Statistics*, Vol. I, P. R. Krishnaiah (ed.) North-Holland, 89–115.

Geisser, S. (1980b). Predictive sample reuse for censored data. In *Bayesian Statistics*, J. M. Bernardo et al. (eds.). Valencia, Spain: University Press.

Geisser, S. (1982). Aspects of the predictive and estimative approaches in the determination of probabilities. *Biometrics* **38**, suppl. 75–93.

Geisser, S. (1984). Predicting Pareto and exponential observables. *Canadian Journal of Statistics* **12**, 143–152.

Geisser, S. (1985a). On the predicting of observables: A selective update. In *Bayesian Statistics 2*, J. M. Bernardo et al. (eds.). Amsterdam: North-Holland, 203–230.

Geisser, S. (1985b). Interval prediction for Pareto and Exponential observables. *Journal of Econometrics* **29**, 173–185.

Geisser, S. (1986). Contribution to discussion. *Journal of the Royal Statistical Society B* **47**, 31–32.

Geisser, S. (1987a). Some remarks on exchangeable normal variables with applications. In *Contributions to the Theory and Application of Statistics*, A. Gelfand (ed.). New York: Academic Press, 127–153.

Geisser, S. (1987b). Influential observations, diagnostics and discordancy tests. *Journal of Applied Statistics* **14**(2), 133–142.

Geisser, S. (1987c). Comments on Gastwirth, the statistical precision of medical screening tests. *Statistical Science*, **2**, 231–232.

Geisser, S. (1988). The future of statistics in retrospect. In *Bayesian Statistics 3*, J. M. Bernardo et al. (eds.). Oxford: Oxford University Press, 147–158.

Geisser, S. (1989). Predictive discordancy tests for exponential observations. *The Canadian Journal of Statistics* **17**(1), 19–26.

Geisser, S. (1990a). On hierarchical Bayes procedures for predicting simple exponential survival. *Biometrics* **46**, 225–230.

Geisser, S. (1990b). Predictive approaches to discordancy testing. In *Bayesian and Likelihood Methods in Statistics and Econometrics*, S. Geisser et al. (eds.). Amsterdam: North-Holland, 321–335.

Geisser, S. (1991). Diagnostics, divergences, and perturbation analysis. In *Directions in Robust Statistics and Diagnostics*, W. Stahel and S. Weisberg, (eds.). Springer-Verlag, 89–100.

Geisser, S. (1992a). Bayesian perturbation diagnostics and robustness. In *Bayesian Analysis in Statistics and Econometrics*, *Lecture Notes in Statistics* 75. Berlin: Springer-Verlag, 289–302.

Geisser, S. (1992b). On the curtailment of sampling. *The Canadian Journal of Statistics* **20**(3), 297–309.

Geisser, S., and Cornfield, J. (1963). Posterior distributions for multivariate normal parameters. *Journal of the Royal Statistical Society B* **25**, 368–376.

Geisser, S., and Desu, M. M. (1968). Predictive zero-mean uniform discrimination. *Biometrika* 55, 519–524.

Geisser, S., and Eddy, W. F. (1979). A predictive approach to model selection. *Journal of the American Statistical Association* 14, 153–160.

Geisser, S., and Johnson, W. (1992). Optimal administration of dual screening tests for detecting a characteristic with special relevance to low prevalence diseases. *Biometrics* 48, 839–852.

Geisser, S., and Johnson, W. (1993a). Interim analysis for normally distributed observables. *Proceedings of the International Symposium on Multivariate Analysis and its Applications* (to appear).

Geisser, S., and Johnson, W. (1993b). Sample size considerations in multivariate normal classification (to appear).

Gelfand, A. E., and Smith, A. F. M. (1990). Sampling based approaches to calculating marginal densities. *Journal of the American Statistical Association* 85, 398–409.

Gnedenko, B. B., Belyavcv, Y. K., and Solovyev, A. D. (1969). *Mathematical Methods of Reliability Theory*. New York: Academic Press.

Grubbs, F. E. (1971). Fiducial bounds on reliability for the two-parameter negative exponential distribution. *Technometrics* 13, 873–876.

Heath, D., and Sudderth, W. (1976). De Finetti's theorem on exchangeable variables. *The American Statistician* 30(4), 188–189.

Hill, B. M. (1968). Posterior distribution of percentiles: Bayes theorem from a finite population. *Journal of the American Statistical Association* 63, 677–691.

Hinkley, D. V. (1979). Predictive likelihood. *Annals of Statistics* 7(4), 718–728.

Hoadley, A. B. (1970). A Bayesian look at inverse linear regression. *Journal of the American Statistical Association* 55, 356–369.

Jeffreys, H. (1946). An invariant form for the prior probability in estimation problems. *Proceedings of the Royal Society of London A* 186, 453–454.

Johnson, W. O., and Gastwirth, J. L. (1991). Bayesian inference for medical screening tests: Approximations useful for the analysis of AIDS data. *Journal of the Royal Statistical Society B* 53(2), 427–440.

Johnson, W., and Geisser, S. (1982). Assessing the predictive influence of observations. In *Statistics and Probability Essays in Honor of C. R. Rao*, G. Kallianpur, P. R. Krishnaiah, and J. K. Ghosh (eds.). Amsterdam: North-Holland, 343–348.

Johnson, W., and Geisser, S. (1983). A predictive view of the detection and characterization of influential observations in regression analysis. *Journal of the American Statistical Association* 78, 137–144.

Kalbfleisch, J. D. (1971). Likelihood methods of prediction. In *Foundations of Statistical Inference*, V. P. Godambe and D. A. Sprott (eds.). New York: Holt, Rinehart & Winston, 378–392.

Kullback, S. (1959). *Information theory and statistics*. New York: Wiley.

Kullback, S., and Leibler, R. A. (1951). On information and sufficiency. *Annals of Mathematical Statistics* 22, 79–86.

Lane, D. A., and Sudderth, W. D. (1984). Coherent predictive inference. *Sankhya: The Indian Journal of Statistics* 46(A, 2), 166–185.

Lauritzen, S. L. (1974). Sufficiency, prediction and extreme models. *Scandinavian Journal of Statistics* 1, 128–134.

Lavine, M. (1987). Prior influence in Bayesian statistics. Unpublished doctoral dissertation. University of Minnesota.

Lee, J. C., and Geisser, S. (1972). Growth curve prediction. *Sankhya A* 34, 393–412.

Lee, J. C., and Geisser, S. (1975). Applications of growth curve prediction. *Sankhya A* 37, 239–256.

Lejeune, M., and Faulkenberry, G. D. (1982). A simple predictive density function. *Journal of the American Statistical Association* 77(379), 654–659.

Leonard, T. (1982). Comment. *Journal of the American Statistical Association* 77(379), 657–658.

Likes, J. (1966). Distribution of Dixon's statistics in the case of an exponential population. *Metrika* 11, 46–54.

Lindley, D. V. (1968). The choice of variables in multiple regression. *Journal of the Royal Statistical Society B* 30, 31–66.

Lindley, D. V., and Barnett, B. N. (1965). Sequential sampling: Two decision problems with linear losses for binomial and normal variables. *Biometrika* 52, 507–532.

Loeve, M. (1960). *Probability Theory*. New York: van Nostrand.

McCulloch, R. (1989). Local model influences. *Journal of the American Statistical Association* 84, 472–478.

Mickey, M. R., Dunn, O. J., and Clark, V. (1967). Note on the use of stepwise regression in detecting outliers. *Biomedical Research* 1, 105–109.

Morris, C. N. (1983). Parametric empirical Bayes inference: theory and applications. *Journal of the American Statistical Association* 78, 47–55.

Murray, G. D. (1977). A note on the estimation of probability density functions. *Biometrika* 64, 150–152.

Nusbacher, J., Chiavetta, J., Naiman, R., Buchner, B., Scalia, V., and Horst, R. (1986). Evaluation of a confidential method of excluding blood donors exposed to human immunodeficiency virus. *Transfusion* 26, 539–541.

Pitman, E. J. G. (1979). *Some Basic Theory for Statistical Influence*. London: Chapman and Hall.

Raiffa, H., and Schlaifer, R. (1961). *Applied Statistical Decision Theory*. Cambridge: Harvard University Press.

Rao, C. R. (1965). *Linear Statistical Inference and its Applications.* New York: Wiley.

Sandstrom, E. G., Schooley, R. T., Ho, D. D., Byington, R., Sarnyadharan, M. G., MacLane, M. E., Essex, M., Gallo, R. C., and Hirsch, M. S. (1985). Detection of anti-HTLV-III antibodies by indirect immunofluorescence using fixed cells. *Transfusion* 25, 308–312.

San Martini, A., and Spezzaferri, F. (1984). A predictive model selection criterion. *Journal of the Royal Statistical Society B* 46(2), 296–303.

Schwarz, Gideon (1978). Estimating the dimension of a model, *Annals of Statistics* 6, 461–464.

Smith, A. F. M., and Spiegelhalter, D. J. (1980). Bayes factors and choice criteria for linear models. *Journal of the Royal Statistical Society B* 42, 213–220.

Stein, C. (1962). Confidence sets for the mean of a multivariate normal distribution. *Journal of the Royal Statistical Society* 24(2), 265–296.

Stigler, S. M. (1982). Thomas Bayes and Bayesian Inference. *Journal of the Royal Statistical Society A* 145(2), 270–258.

Stone, M. (1974). Cross-validatory choice and assessment of statistical predictions. *Journal of the Royal Statistical Society B* 36, 111–147.

Stone, M. (1977). An asymptotic equivalence of choice of model by cross-validation and Akaike's criterion. *Journal of the Royal Statistical Society B* 39, 44–47.

Varian, H. R. (1975). A Bayesian approach to real estate assessment. In *Studies in Bayesian Econometrics and Statistics*, S. Fienberg and A. Zellner (eds.). Amsterdam: North-Holland, 195–208.

West, M., and Harrison, J. (1989). *Bayesian Forecasting and Dynamic Models.* Berlin: Springer.

Zellner, A. (1971). *An Introduction to Bayesian Inference Econometrics.* New York: Wiley.

Zellner, A. (1977). Maximal data information prior distributions. In *New Developments in the Application of Bayesian Methods*, A. Aykac and C. Brumat (eds.). Amsterdam: North-Holland.

Zellner, A. (1986). Bayesian estimation and prediction using asymmetric loss functions. *Journal of the American Statistical Association* 81, 446–451.

Zellner, A., and Chetty, V. K. (1965). Prediction and decision problems in regression models from the Bayesian point of view. *Journal of the American Statistical Association* 60, 608–616.

Author Index

Subject Index